ESSENTIAL SQA EXAM PRACTICE

D1382060

NATIONAL 5 PHYSICS

Practice Questions & Exam Papers

QUESTIONS & PAPERS

Practise **105+ questions** covering every question type and topic

Complete **2 practice papers** that mirror the real SQA exams

Paul Van der Boon

HODDER GIBSON
AN HACHETTE UK COMPANY

The Publishers would like to thank the following for permission to reproduce copyright material.

Photo credits

p.5 Shcherbakov Ilya/Shutterstock; **p.32** eugenesergeev/Fotolia; **p.39** Seksan Kingwatcharapong/Shutterstock; **p.53** (top) Joggie Botma/Shutterstock; **p.53** (bottom) and **p.105** Alexander Gordeyev/Shutterstock.com; **p.55** (left) Kim Steele/Photodisc/Getty Images/Science, Technology & Medicine; **p.55** (right) StockTrek/Photodisc/Getty Images; **p.81** Imagestate Media (John Foxx)/Global Travel Transport Vol. 29; **p.82** dell/Fotolia; **p.83** gary718/123RF; **p.84** (top) Georgios Kollidas/Fotolia; **p.84** (bottom) Morphart Creation/Shutterstock; **p.109** Skierx/Fotolia; **p.112** (left) Brian Jackson/Fotolia; **p.112** (right) Lars Johansson/Fotolia; **p.115** jupeart/Shutterstock; **p121** (top) Rich Carey/Shutterstock; **p.128** Kris Christiaens/123RF.

Acknowledgements

Every effort has been made to trace all copyright holders, but if any have been inadvertently overlooked, the Publishers will be pleased to make the necessary arrangements at the first opportunity.

Although every effort has been made to ensure that website addresses are correct at time of going to press, Hodder Gibson cannot be held responsible for the content of any website mentioned in this book. It is sometimes possible to find a relocated web page by typing in the address of the home page for a website in the URL window of your browser.

Hachette UK's policy is to use papers that are natural, renewable and recyclable products and made from wood grown in well-managed forests and other controlled sources. The logging and manufacturing processes are expected to conform to the environmental regulations of the country of origin.

Orders: please contact Bookpoint Ltd, 130 Park Drive, Milton Park, Abingdon, Oxon OX14 4SE. Telephone: (44) 01235 827827. Fax: (44) 01235 400401. Email education@bookpoint.co.uk. Lines are open from 9 a.m. to 5 p.m., Monday to Friday, with a 24-hour message answering service. Visit our website at www.hoddereducation.co.uk. If you have queries or questions that aren't about an order you can contact us at hoddergibson@hodder.co.uk.

© Paul Van der Boon 2019

First published in 2019 by

Hodder Gibson, an imprint of Hodder Education

An Hachette UK Company

211 St Vincent Street

Glasgow, G2 5QY

Impression number	5	4	3	2	1
Year	2023	2022	2021	2020	2019

Illustrations by Aptara Inc.

Typeset in India by Aptara Inc.

Printed and bound by CPI Group (UK) Ltd, Croydon CR0 4YY

A catalogue record for this title is available from the British Library.

ISBN: 978 1 5104 7191 7

MIX
Paper from
responsible sources
FSC™ C104740
www.fsc.org

SCOTLAND EXCEL

We are an approved supplier on the Scotland Excel framework.

Schools can find us on their procurement system as:

Hodder & Stoughton Limited t/a Hodder Gibson.

CONTENTS

INTRODUCTION

National 5 Physics

The assessment materials included in this book are designed to provide practice and to support revision for the National 5 Physics course assessment question paper (the examination), which is worth 80% of the final grade for this course.

The materials are provided in two sections:

1 Practice Questions
2 Practice Papers: Practice Paper 1 and Practice Paper 2.

Practice Questions

The Practice Questions are arranged in sets to reflect the different types of question which are used in the National 5 Physics examination, as shown in the grid below.

There are 325 marks' worth of questions in this section.

A Multiple choice (1 mark)	B Course content		C Experimental and data handling	D Open ended	E Scientific literacy
	B1: Short	B2: Extended			
Objective questions testing knowledge and understanding (KU) within Key Areas and skills	Short questions testing course content KU within Key Areas – in the exam these will be part of a multi-part 'Extended question'	Structured extended exam-type questions testing course content KU within Key Areas	Structured questions testing data-handling and experimental skills	Questions requiring you to use your knowledge of physics to give an extended comment on a physics phenomenon or situation in an ordered way	Structured questions testing certain skills and KU based on a piece of science writing, sometimes requiring the use of an unfamiliar equation

Linking questions to course specification

National 5 Physics is split into six areas of knowledge and understanding: Dynamics, Space, Electricity, Properties of matter, Waves and Radiation. Each of these areas is further split into Key Areas. These are numbered in the margin, so that as you work through the book you can identify which Key Areas you are scoring well in and which Key Areas you need to focus your revision on further (see Practice Questions Key Area index grid).

▶ Dynamics:
- D1 vectors and scalars
- D2 velocity–time graphs
- D3 acceleration
- D4 Newton's laws
- D5 energy
- D6 projectile motion

▶ Space:
- S1 space exploration
- S2 cosmology

▶ Electricity:
- E1 electrical charge carriers
- E2 potential difference (voltage)
- E3 Ohm's law
- E4 practical electrical and electronic circuits
- E5 electrical power

▶ Properties of matter:
- PM1 specific heat capacity
- PM2 specific latent heat
- PM3 gas laws and the kinetic model

▶ Waves:
- W1 wave parameters and behaviours
- W2 electromagnetic spectrum
- W3 refraction of light

▶ Radiation:
- R1 nuclear radiation

Key Area index grid

The **Practice Questions Key Area index grid** on page vi shows the pattern of coverage of the Key Areas and the skills across the Practice Questions.

After working on questions from Key Areas across an area of study, you might want to use the boxes to assess your progress. We suggest marking like this [–] if you are having difficulty (less than half marks), like this [+] if you have done further work and are more comfortable (more than half marks) and this [*] if you are confident you have learned and understood an entire area of study (nearly full marks). Alternatively, you could use a 'traffic light' system using colours – red for 'not understood'; orange for 'more work needed' and green for 'fully understood'. **If you continue to struggle with a set of Key Area questions, you should see your teacher for extra help**.

Practice Papers

This book contains two full practice papers worth 135 marks each. These papers are designed to be similar to the actual paper you will sit. They will generally follow the same order, have the same structure and will contain questions of the type that you will encounter in your examination.

Section 1 – This is an objective test, which contains 25 multiple-choice items worth 1 mark each and totalling 25 marks altogether.

Section 2 – This contains restricted and extended-response questions worth around 3 to 12 marks each, totalling 110 marks altogether.

In each paper, the marks are distributed proportionately across all six topics of the course content and about 90–100 marks are for the demonstration and application of knowledge. The remaining 35–45 marks are for the application of skills of scientific inquiry.

70% of the marks in each paper are set at the standard of Grade C and the remaining 30% are more difficult marks set at the standard for Grade A.

Grading

Each paper has a total of 135 marks – if you score 67 marks that's a C pass. You will need about 80 marks for a B pass and about 95 marks for an A. These figures are a rough guide only.

Timing

If you are attempting a full paper in this book, limit yourself to **2 and a half hours** to complete it. Get someone to time you! We recommend no more than 30 minutes for Section 1 and the remainder of the time for Section 2.

Key Area index grid

The **Practice Papers Key Area index grid** on page vii shows the pattern of coverage of the Key Areas and the skills across the practice papers. This can be used in the same way as the Practice Questions Key Areas index grid (see earlier) to aid your revision.

Using the questions and papers

We recommend working between attempting questions or papers and studying the answers (see below).

Where any difficulty is encountered, it's worth trying to consolidate your knowledge and skills. Use the information in the Student margin to identify the type of question you find trickiest.

You will need a **pen**, a **sharp pencil**, a **clear plastic ruler** and a **calculator** for the best results. A couple of different **coloured highlighters** could also be handy.

Answers

The answers for the Practice Questions are provided on pages 60–72 and those for the Practice Papers on pages 129–148. They give National Standard answers but, occasionally, there may be other acceptable answers.

The answers to the practice papers (Section 2) also have commentaries with hints and tips provided alongside. Don't feel you need to use them all!

The commentaries in the answers focus on the correct physics, as well as hints, advice on wording of answers and notes of commonly made errors.

Revision

There are 20 Key Areas from all six topics of the course content, so covering two each week would require about a 10-week revision programme. The exams are in May, so starting after your February holiday would give you time. You could use the Revision Calendar at www.hoddergibson.co.uk/ESEP-extras to help you plan and keep a record of your progress.

We wish you the very best of luck!

KEY AREA INDEX GRIDS

Practice Questions

Area	Key Area	A Multiple choice (1 mark)	B Course content		C Experimental and data handling	D Open ended	E Scientific literacy	Check
			Short	Extended				
Dynamics	D1 Vectors and scalars	1–3	1–10	1–3	1–2	1–2		88
Dynamics	D2 Velocity–time graphs	4	1–10	1–3	1–2	1–2		88
Dynamics	D3 Acceleration	5	1–10	1–3	1–2	1–2		88
Dynamics	D4 Newton's laws	6–9	1–10	1–3	1–2	1–2		88
Dynamics	D5 Energy	10–13	1–10	1–3	1–2	1–2		88
Dynamics	D6 Projectile motion	14	1–10	1–3	1–2	1–2		88
Space	S1 Space exploration	15–16	11–15	4	3	3	1–2	50
Space	S2 Cosmology	17	11–15	4	3	3	1–2	50
Electricity	E1 Electrical charge carriers	18	16–21	5–6	4–5			59
Electricity	E2 Potential difference (voltage)	19–20	16–21	5–6	4–5			59
Electricity	E3 Ohm's law	21–22	16–21	5–6	4–5			59
Electricity	E4 Practical electrical and electronic circuits	23–24	16–21	5–6	4–5			59
Electricity	E5 Electrical power	25	16–21	5–6	4–5			59
Properties of matter	PM1 Specific heat capacity	26	22–25	7–8	6	4–5		48
Properties of matter	PM2 Specific latent heat	27–28	22–25	7–8	6	4–5		48
Properties of matter	PM3 Gas laws and the kinetic model	29–33	22–25	7–8	6	4–5		48
Waves	W1 Wave parameters and behaviours	34–35	26–30	9–11	7	6		41
Waves	W2 Electromagnetic spectrum	36	26–30	9–11	7	6		41
Waves	W3 Refraction of light	37	26–30	9–11	7	6		41
Radiation	R1 Nuclear radiation	38–44	31–34	12	8			34
Totals		44	114	92	42	18	10	320

Practice Paper 1

Area	Key Area	Section 1	Section 2					Check
		Multiple choice	Course content – extended		Experimental and data handling	Open ended	Scientific literacy	
			Calculate/ show that	State/ explain/ describe				
Dynamics	D1 Vectors and scalars	1, 2					4c	5
Dynamics	D2 Velocity–time graphs		1aiii					3
Dynamics	D3 Acceleration	3			1ai			4
Dynamics	D4 Newton's laws	4	1aii, 1b			3		11
Dynamics	D5 Energy		2 a, 2 b, 5 ai	2c				10
Dynamics	D6 Projectile motion	5						1
Space	S1 Space exploration	6,7	5 bi, 5 bii	5 aii, 5 aiii			4a	11
Space	S2 Cosmology	8	12b					4
Electricity	E1 Electrical charge carriers	9, 10						2
Electricity	E2 Potential difference (voltage)	11						1
Electricity	E3 Ohm's law	12	6 aiii					5
Electricity	E4 Practical electrical and electronic circuits	13, 14	7 aii, 7 bi	6 aii, 6 b, 7 bii	6 ai			12
Electricity	E5 Electrical power		7 ai, 8 b					6

Area	Key Area	Section 1	Section 2					Check
		Multiple choice	Course content – extended		Experimental and data handling	Open ended	Scientific literacy	
			Calculate/ show that	State/ explain/ describe				
Properties of matter	PM1 Specific heat capacity				8a			☐ 3
	PM2 Specific latent heat	15		9a, 9c	9b			☐ 7
	PM3 Gas laws and the kinetic model	16	10 a, 10 bi	10 bii				☐ 8
Waves	W1 Wave parameters and behaviours	17	11ai, 11aii, 12cii	12 ci, 11 aiii				☐ 13
	W2 Electromagnetic spectrum			12 ai, 12 aii, 12 aiii		13	4b	☐ 8
	W3 Refraction of light	18		11b				☐ 4
Radiation	R1 Nuclear radiation	19–25	1c, 14 cii	14 bi, 14 ci	14 a, 14 bii			☐ 17
Totals		25	62	23	13	6	6	/135

Practice Paper 2

Area	Key Area	Section 1	Section 2					Check
		Multiple choice	Course content – extended		Experimental and data handling	Open ended	Scientific literacy	
			Calculate/ show that	State/ explain/ describe				
Dynamics	D1 Vectors and scalars	1	3 biii, 5 c	10 biii				9
	D2 Velocity–time graphs	2						1
	D3 Acceleration	3, 4						2
	D4 Newton's laws	5, 6	1b	1c	3ai			8
	D5 Energy	7	3aii, 3bii	8biii		2		11
	D6 Projectile motion	8	3bi					4
Space	S1 Space exploration	9, 10, 11, 12		5aii, 5d			4a	8
	S2 Cosmology	13					4b, c	5
Electricity	E1 Electrical charge carriers	14		11a				2
	E2 Potential difference (voltage)			1a				1
	E3 Ohm's law	15, 16			6 bi, 6 bii, 6 biii			9
	E4 Practical electrical and electronic circuits	17, 18		6a, 7ai, 11ci, 11cii				6
	E5 Electrical power	19	7aii, 7 b, 8 bi					10

Area	Key Area	Section 1	Section 2					Check
		Multiple choice	Course content – extended		Experimental and data handling	Open ended	Scientific literacy	
			Calculate/ show that	State/ explain/ describe				
Properties of matter	PM1 Specific heat capacity		8a					3
	PM2 Specific latent heat		8 bii					3
	PM3 Gas laws and the kinetic model	20, 21	9 b, 10 a, 10 bi, 10 bii	9 a				13
Waves	W1 Wave parameters and behaviours	22	5b, 11biii		11 bi, 11bii	12		12
	W2 Electromagnetic spectrum			5 ai			4d	3
	W3 Refraction of light	23, 24						2
Radiation	R1 Nuclear radiation	25	13f, 14bii, 14ci, 14cii	13a–e, 14a, 14ciii, 14d	14bi			23
Totals		25	61	26	11	6	6	/135

RELATIONSHIPS SHEET

You will be provided with the Relationships sheet in your final exam.
Please refer to it as required for each Practice Paper.

$d = vt$

$d = \bar{v}t$

$s = vt$

$s = \bar{v}t$

$a = \dfrac{v - u}{t}$

$F = ma$

$W = mg$

$E_w = Fd$

$E_p = mgh$

$E_k = \dfrac{1}{2}mv^2$

$Q = It$

$V = IR$

$V_2 = \left(\dfrac{R_2}{R_1 + R_2}\right)V_s$

$\dfrac{V_1}{V_2} = \dfrac{R_1}{R_2}$

$R_T = R_1 + R_2 + \dots$

$\dfrac{1}{R_T} = \dfrac{1}{R_1} + \dfrac{1}{R_2} + \dots$

$P = \dfrac{E}{t}$

$P = IV$

$P = I^2 R$

$P = \dfrac{V^2}{R}$

$E_h = cm\,\Delta T$

$E_h = ml$

$p = \dfrac{F}{A}$

$p_1 V_1 = p_2 V_2$

$\dfrac{p_1}{T_1} = \dfrac{p_2}{T_2}$

$\dfrac{V_1}{T_1} = \dfrac{V_2}{T_2}$

$\dfrac{pV}{T} = \text{constant}$

$f = \dfrac{N}{t}$

$v = f\lambda$

$T = \dfrac{1}{f}$

$A = \dfrac{N}{t}$

$D = \dfrac{E}{m}$

$H = Dw_r$

$\dot{H} = \dfrac{H}{t}$

Additional relationships

Circle

circumference $= 2\pi r$

area $= \pi r^2$

Sphere

area $= 4\pi r^2$

volume $= \dfrac{4}{3}\pi r^3$

Trigonometry

$\sin\theta = \dfrac{\text{opposite}}{\text{hypotenuse}}$

$\cos\theta = \dfrac{\text{adjacent}}{\text{hypotenuse}}$

$\tan\theta = \dfrac{\text{opposite}}{\text{adjacent}}$

$\sin^2\theta + \cos^2\theta = 1$

DATA SHEET

Speed of light in materials

Material	Speed in m s^{-1}
Air	3.0×10^8
Carbon dioxide	3.0×10^8
Diamond	1.2×10^8
Glass	2.0×10^8
Glycerol	2.1×10^8
Water	2.3×10^8

Gravitational field strengths

	Gravitational field strength on the surface in N kg^{-1}
Earth	9.8
Jupiter	23
Mars	3.7
Mercury	3.7
Moon	1.6
Neptune	11
Saturn	9.0
Sun	270
Uranus	8.7
Venus	8.9

Specific latent heat of fusion of materials

Material	Specific latent heat of fusion in J kg^{-1}
Alcohol	0.99×10^5
Aluminium	3.95×10^5
Carbon Dioxide	1.80×10^5
Copper	2.05×10^5
Iron	2.67×10^5
Lead	0.25×10^5
Water	3.34×10^5

Specific latent heat of vaporisation of materials

Material	Specific latent heat of vaporisation in J kg^{-1}
Alcohol	11.2×10^5
Carbon Dioxide	3.77×10^5
Glycerol	8.30×10^5
Turpentine	2.90×10^5
Water	22.6×10^5

Speed of sound in materials

Material	Speed in m s^{-1}
Aluminium	5200
Air	340
Bone	4100
Carbon dioxide	270
Glycerol	1900
Muscle	1600
Steel	5200
Tissue	1500
Water	1500

Specific heat capacity of materials

Material	Specific heat capacity in J kg^{-1} °C^{-1}
Alcohol	2350
Aluminium	902
Copper	386
Glass	500
Ice	2100
Iron	480
Lead	128
Oil	2130
Water	4180

Melting and boiling points of materials

Material	Melting point in °C	Boiling point in °C
Alcohol	−98	65
Aluminium	660	2470
Copper	1077	2567
Glycerol	18	290
Lead	328	1737
Iron	1537	2737

Radiation weighting factors

Type of radiation	Radiation weighting factor
alpha	20
beta	1
fast neutrons	10
gamma	1
slow neutrons	3

Question type: Multiple-choice

>> HOW TO ANSWER

In the National 5 Physics exam, **Section 1** contains 25 multiple-choice questions worth **1 mark** each and totalling 25 marks altogether. Each question has five possible choices of answer.

Only one answer is correct.

The multiple-choice questions are distributed proportionally across all six areas of the course content and are designed to test a range of skills, for example:

▶ knowledge and understanding of the course

▶ using equations

▶ selecting correct statements from a list

▶ selecting and analysing information from a diagram.

One question on average should take just over a minute (i.e. 67 seconds). In practice, some questions might take a bit longer, for example, if there is a lot to read or if calculations or other information processing are involved, but others can be answered more quickly if it's just straightforward recall.

It is important to practise as many multiple-choice questions as possible to get used to the 'style' and types of questions.

Where a question is complicated, write down notes and working on the blank pages at the end of the question paper or beside the actual question. **Do not use the answer grid for working** and remember to cross out any rough working for these multiple-choice questions when you have finished.

Top Tip!

Be aware of how much time you spend on each question. For example, **do not** spend 5 minutes on one question worth only 1 mark, especially if you haven't completed the rest of the questions – you can always return to the question later if there's time.

Top Tip!

You should spend no more than 30 minutes on Section 1 in your examination.

---

1 Which of the following contains one vector and one scalar quantity?

   A time, energy

   B force, mass

   C acceleration, displacement

   D speed, time

   E velocity, force

STUDENT MARGIN

Dynamics 1

2 A cyclist follows the route shown in the diagram.

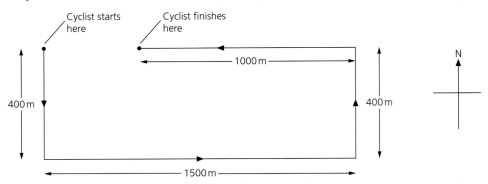

Which row in the table shows the total distance cycled and the final displacement of the cyclist from start to finish?

	Total distance	Final displacement
A	3300 m	500 m East
B	3300 m	1000 m East
C	3800 m	1000 m West
D	3800 m	500 m East
E	3300 m	1000 m West

3 A cyclist follows a course from P to R as shown.

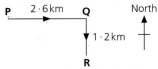

The total journey takes 1 hour.

Which row in the table gives the average speed and the average velocity of the vehicle for the whole journey?

	Average speed	Average velocity
A	2·9 km h^{-1}	3·8 km h^{-1} (115)
B	2·9 km h^{-1} (025)	3·8 km h^{-1}
C	3·8 km h^{-1}	2·9 km h^{-1} (115)
D	3·8 km h^{-1}	2·9 km h^{-1} (025)
E	3·8 km h^{-1} (115)	2·9 km h^{-1}

$2·6^2 + 3·2^2 = 8·2$

$\sqrt{8·2^2} = 2·9$

speed = distance ÷ time

4 As a car approaches a town, the driver applies the brakes. The speed–time graph of the car's motion is shown.

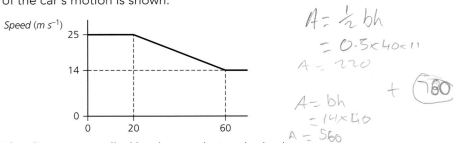

$A = \frac{1}{2} bh$
$= 0.5 \times 40 \times 11$
$A = 220$

$A = bh$
$= 14 \times 40$
$A = 560$

$+ \boxed{780}$

The distance travelled by the car during the braking is

A 560 m

B 780 m

C 1000 m

D 1060 m

E 1500 m.

Dynamics 3

5 A child sledges down a hill.

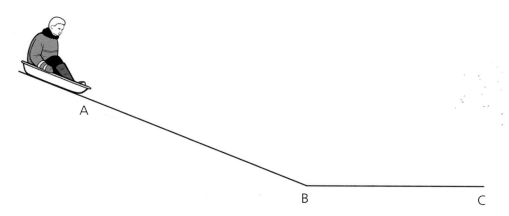

The sledge and child are released from rest at point A. They reach a speed of $4 \cdot 5 \, \text{m s}^{-1}$ at point B.

The sledge and child take 9 s to reach point B.

The acceleration of the sledge, in m s^{-2}, is

A   −0·5

B   (0·5)

C   2·0

D   13·5

E   40·5.

$$a = \frac{v - u}{t}$$

$$a = \frac{4 \cdot 5 - 0}{9}$$

$$a = 0 \cdot 5 \, \text{m o}^{-2}$$

6 The engine in a speedboat provides a constant forward thrust on the speedboat.

The force of friction between the boat and the water increases as the speed of the boat increases.

The boat starts from rest.

Which of the graphs shows how the speed of the boat varies with time?

Dynamics 4

A

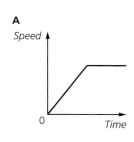

B

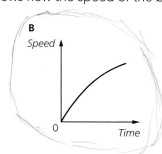

C

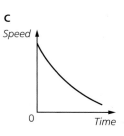

D

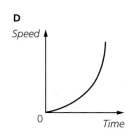

E

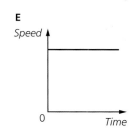

7 A constant force of 54 N pulls a wooden block of mass 5 kg on a track. The block is moving in the direction shown.

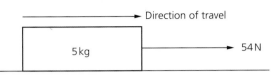

At one instant, the block is **decelerating** at $1.2\,\text{m}\,\text{s}^{-2}$.

The force of friction acting on the block at this instant is

A   zero

B   48 N

C   54 N

D   60 N

E   114 N.

Dynamics 4

8 An aircraft engine exerts a force on the air.

Which of the following completes the 'Newton pair' of forces?

A   The force of friction exerts a force between the air and the engine.

B   The Earth exerts a force on the engine.

C   The force of friction exerts a force between the engine and the air.

D   The engine exerts a force on the Earth.

E   The air exerts a force on the engine.

Dynamics 4

9 An equation known as Stokes' law can be used to determine the upward drag force, F_d, acting on a small drop of oil falling under gravity through air when it has reached its terminal velocity:

$F_d = 6\pi r \eta v_1$

where:

v_1 is the terminal velocity of the falling drop

η is the viscosity of the air

r is the radius of the drop.

The values for one particular oil drop were:

radius: $2.83 \times 10^{-6}\,\text{m}$

terminal velocity: $8.56 \times 10^{-4}\,\text{m}\,\text{s}^{-1}$

viscosity of air: $1.820 \times 10^{-5}\,\text{kg}\,\text{m}^{-1}\,\text{s}^{-1}$

The value of the drag force, F_d, acting on the oil drop is

A $4.40 \times 10^{-14}\,\text{N}$

B $8.31 \times 10^{-13}\,\text{N}$

C $9.70 \times 10^{-10}\,\text{N}$

D $4.57 \times 10^{-8}\,\text{N}$

E $2.90 \times 10^{-8}\,\text{N}$.

Dynamics 4

10 A builder pushes a wheelbarrow of total mass 50 kg along a path with a force of 20 N at a constant speed of 4·0 m s⁻¹ for a distance of 120 m.

The force of friction acting on the wheelbarrow is 20 N.

The work done on the wheelbarrow by the builder is

A   0 J

B   400 J

C   490 J

D   2400 J

E   9600 J.

> Dynamics 5

11 The total mass of a go-cart and driver is 200 kg. While braking, the cart and driver are brought to rest from a speed of 10 m s⁻¹ in a time of 8·0 s.

The maximum energy that could be transformed into heat in the brakes is

A   2000 J

B   10 000 J

C   16 000 J

D   20 000 J

E   40 000 J.

> Dynamics 5

12 A ball is dropped vertically from a height of 1·20 m to the ground and rebounds as shown.

The mass of the ball is 0·40 kg.

After the rebound the ball reaches a height 0·25 m lower than the release height.

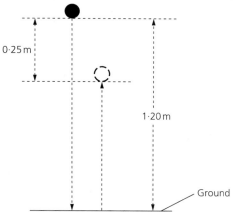

The amount of energy lost is

A   0·10 J

B   0·15 J

C   0·98 J

D   3·7 J

E   4·7 J.

> Dynamics 5

13 A ball of mass 0·75kg is released from a height of 2·0m and falls towards the ground.

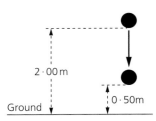

2·00m

0·50m

Ground

Which row in the table shows the gravitational potential energy and the kinetic energy of the ball when it is at a height of 0·50 m from the floor?

	Gravitational potential energy (J)	Kinetic energy (J)
A	3·7	3·7
B	3·7	11·0
C	11·0	3·7
D	14·0	3·7
E	14·0	11·0

14 A package is released from a helicopter flying horizontally at a constant speed of $30\,\mathrm{m\,s^{-1}}$.

Which diagram represents the path of the package after its release until it reaches the ground?

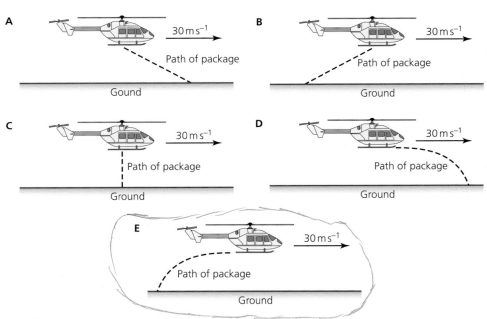

15 The altitude above the Earth's surface of a geostationary satellite is approximately

A   360 km

B   3600 km

C   36 000 km

D   360 000 km

E   3 600 000 km.

Dynamics 5

Dynamics 6

Space 1

16 At launch, an Apollo lunar ascent module accelerated vertically upward from the surface of the Moon at $1.4\,\text{m}\,\text{s}^{-2}$. The mass of the module was $5000\,\text{kg}$.

What was the size of the upward thrust acting on the ascent module at this time?

A 7000 N

B 8000 N

C 15 000 N

D 49 000 N

E 56 000 N

Space 1

17 A star is 960 light years from the Earth.

The distance of the star from the Earth, in metres, is

A 3.0×10^{10}

B 1.1×10^{14}

C 9.5×10^{15}

D 1.5×10^{17}

E 9.1×10^{18}.

Space 2

18 The current in a $16\,\Omega$ resistor is $4\,\text{A}$.

The charge passing through the resistor in 8 seconds is

A 2 C

B 4 C

C 8 C

D 32 C

E 128 C.

Electricity 1

19 Atomic particles electrons, protons and neutrons are directed into an electric field as shown in the diagram.

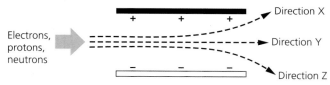

Electrons, protons, neutrons

Direction X

Direction Y

Direction Z

Which row in the table correctly identifies how the particles are deflected?

	Direction X	Direction Y	Direction Z
A	electrons	protons	neutrons
B	protons	electrons	neutrons
C	neutrons	electrons	protons
D	neutrons	protons	electrons
E	electrons	neutrons	protons

Electricity 2

20 An electrical circuit has a voltage supply.

Voltage is a measure of the

A current in the circuit

B energy given to the charges in the circuit

C power developed in the circuit

D total resistance of the circuit

E number of charges in the circuit.

Electricity 2

21 The circuit below controls the brightness of two identical 3·0 V lamps.

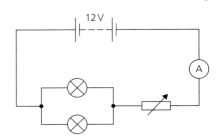

When the lamps are operating at their correct voltage, the ammeter reading is 0·75 A.

The resistance of the variable resistor is

A   4·0 Ω

B   6·75 Ω

C   8·0 Ω

D   12·0 Ω

E   16·0 Ω.

22 The circuit shows two resistors connected to a supply voltage.

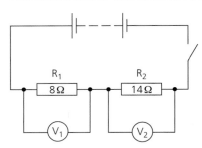

When switched on, the reading on voltmeter V_2 is 5·5 V.

The reading on voltmeter V_1 is

A   0·1 V

B   3·1 V

C   9·6 V

D   121 V

E   616 V.

23 Resistors are connected in the following circuit as shown.

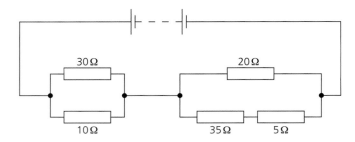

In which resistor is the current **greatest**?

A   5 Ω

B   10 Ω

C   20 Ω

D   30 Ω

E   35 Ω.

24 Four circuit symbols, P, Q, R and S are shown.

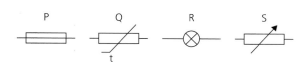

Which row in the table identifies the components represented by these symbols?

	P	Q	R	S
A	variable resistor	LDR	battery	thermistor
B	variable resistor	fuse	lamp	LDR
C	fuse	thermistor	lamp	variable resistor
D	resistor	variable resistor	lamp	fuse
E	LDR	fuse	resistor	thermistor

Electricity 4

25 The information shown is for an electric heater.

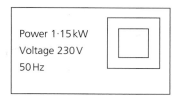

Power 1·15 kW
Voltage 230 V
50 Hz

The resistance of the heater is

A   5 Ω
B   46 Ω
C   264·5 Ω
D   46 000 Ω
E   264 500 Ω.

Electricity 5

26 A mass of 2·5 kg of water is heated up in a flask. A heater supplies $6·27 \times 10^5$ J of heat energy to the water.

The maximum temperature change of the water is

A   60·0 °C
B   120·0 °C
C   160·0 °C
D   320·0 °C
E   640·0 °C.

$$\Delta = \frac{E_h}{c\,m}$$

$$= \frac{6.27 \times 10^5}{4180 \times 2.5}$$

$$\Delta T = 60$$

Properties of matter 1

27 A solid substance is placed in an insulated container and heated continuously with an immersion heater.

The graph shows how the temperature of the substance in the container changes with time.

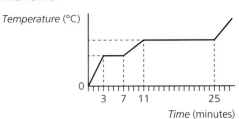

After 25 minutes the substance is a

A   mixture of solid and liquid

B   mixture of liquid and gas

C   solid

D   liquid

E   gas.

Properties of matter 2

28 A sample of water is at a temperature of 0 °C.

9.29×10^5 J of heat energy is removed from the sample.

The mass of water changed into ice is

A   0.36 kg

B   0.41 kg

C   2.43 kg

D   2.78 kg

E 3.11×10^{11} kg.

Properties of matter 2

29 The outside pressure on a submarine beneath the sea is 3.0×10^6 Pa. The air pressure inside the submarine is 1.0×10^5 Pa.

The area of a door hatch on the submarine is $1.8 \, m^2$.

What is the inward force on the door hatch due to the pressure difference?

A 1.61×10^6 N

B 1.72×10^6 N

C 5.22×10^6 N

D 5.40×10^6 N

E 5.58×10^6 N.

Properties of matter 3

30 The mean kinetic energy of the particles in a substance is a measure of its

A   activity.

B   temperature

C   pressure

D   volume

E   density.

Properties of matter 3

31 A solid is cooled from 75 °C to 49 °C. The temperature change in kelvin is

A   26 K

B   124 K

C   299 K

D   397 K

E   572 K.

Properties of matter 3

32 A student investigates the relationship between the pressure and kelvin temperature of a fixed mass of gas at constant volume.

Which graph correctly shows the relationship?

A

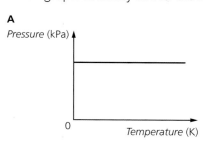

B

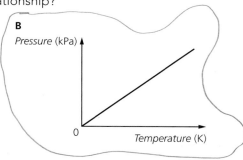

C

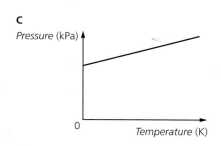

D

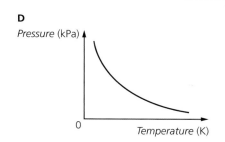

E

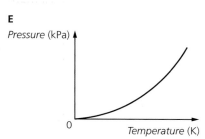

33 The pressure of a fixed mass of gas is $5·0 \times 10^5$ Pa.

The temperature of the gas is 37 °C and the volume of the gas is $3·5\,m^3$.

The temperature of the gas increases to 66 °C and the volume of the gas increases to $6·0\,m^3$.

What is the new pressure of the gas?

A   100 Pa

B $2·7 \times 10^5$ Pa

C $3·2 \times 10^5$ Pa

D $5·2 \times 10^5$ Pa

E $2·5 \times 10^9$ Pa.

34 The period of vibration of a violin string is 4 ms.

The frequency of the sound produced by the violin string is

A   0·25 Hz

B   25 Hz

C   250 Hz

D   400 Hz

E   4000 Hz.

Properties of matter 3

Properties of matter 3

Waves 1

35 Water waves are produced in a ripple tank. The waves reach a barrier and are diffracted.

Which diagram correctly shows the diffraction in the ripple tank?

Waves 1

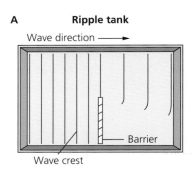

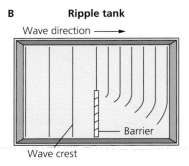

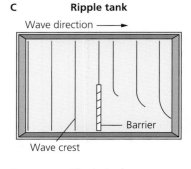

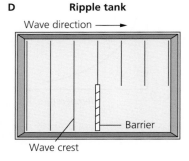

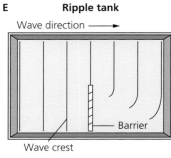

36 A student makes the following statements about ultraviolet rays and X-rays.

I In air, X-rays travel faster than ultraviolet rays.

II In air, X-rays have a longer wavelength than ultraviolet rays.

III X-rays and ultraviolet rays are both members of the electromagnetic spectrum.

Which of these statements is/are correct?

A I only

B III only

C I and II only

D I and III only

E II and III only.

Waves 2

37 The diagram shows the path of a ray of green light as it passes from a glass block into air.

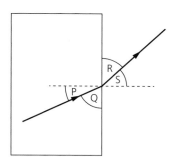

Which row in the table shows the angle of incidence and the angle of refraction?

	Angle of incidence	Angle of refraction
A	Q	R
B	R	Q
C	P	S
D	S	P
E	P	R

38 A nuclear fission process produced radiation Z, which passed through an electric field as shown.

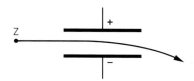

Radiation Z is

A a slow neutron

B a fast neutron

C a beta particle

D an alpha particle

E gamma radiation.

39 Which row in the table correctly describes alpha (α) and beta (β) radiations?

	α	β
A	helium nucleus emitted during the radioactive decay of an atom	electron emitted during the radioactive decay of an atom
B	helium nucleus emitted during the radioactive decay of an atom	electromagnetic radiation emitted during the radioactive decay of an atom
C	electromagnetic radiation emitted during radioactive decay of an atom	electron emitted during the radioactive decay of an atom
D	electron emitted during the radioactive decay of an atom	helium nucleus emitted during the radioactive decay of an atom
E	electromagnetic radiation emitted during radioactive decay of an atom	helium nucleus emitted during the radioactive decay of an atom

Margin notes:
- Waves 3
- Radiation 1
- Radiation 1

40 A ratemeter was used to measure the background count rate in a laboratory.

The background count rate was measured as 4 counts per second.

The ratemeter was then used to take measurements of the count rate from a radioactive source.

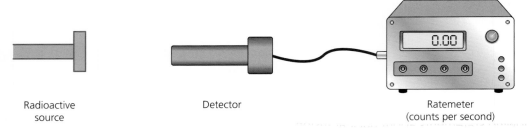

Radioactive source

Detector

Ratemeter (counts per second)

Absorbing materials were placed in turn between the source and detector and the results are given in the table.

Absorbing material	Count rate (counts per second)
No material	75
Sheet of paper	75
3 mm of aluminium	40
10 mm of lead	8

The radiation emitted by the source is

A α only

B γ only

C α and β only

D β and γ only

E α, β and γ.

41 In 30 minutes, $15\cdot84 \times 10^{18}$ nuclei of a radioactive isotope decay.

The activity of the isotope is

A $1\cdot47 \times 10^{14}$ Bq

B $8\cdot80 \times 10^{15}$ Bq

C $5\cdot28 \times 10^{17}$ Bq

D $4\cdot75 \times 10^{20}$ Bq

E $2\cdot85 \times 10^{22}$ Bq.

42 A tissue sample has a mass of 0·08 kg. The tissue receives an absorbed dose of 24 μGy from alpha particles in 3 minutes.

The energy absorbed by the tissue is

A 1·92 μJ

B 5·76 μJ

C 24·0 μJ

D 300 μJ

E 3000 μJ.

Radiation 1

Radiation 1

Radiation 1

43 The following statements refer to nuclear fission and fusion.

I   During nuclear fusion a large nucleus splits into smaller nuclei.

II During nuclear fission neutrons are emitted.

III During nuclear fission and nuclear fusion energy is released.

Which of the statements is/are correct?

A   I only

B   II only

C   III only

D   I and II only

E   II and III only.

| Radiation 1 |

44 A student makes the following statements about nuclear fusion.

I   During nuclear fusion, two nuclei combine to form a nucleus of smaller total mass.

II During nuclear fusion, a large amount of energy is released.

III On Earth, extreme temperatures are required to produce the same nuclear fusion reaction that occurs in the Sun.

Which of these statements is/are correct?

A   I only

B   II only

C   III only

D   I and II only

E   I, II and III.

| Radiation 1 |

Question type: Short

>> HOW TO ANSWER

These short revision questions focus on the techniques and skills required for questions that feature regularly in every exam, such as:

▶ performing calculations using relationship equations
▶ data handling
▶ graph interpretation.

Using equations

More than half of the total marks awarded in Section 2 of the exam are for being able to calculate answers using an equation (relationship) from the Relationship sheet which is supplied with the exam paper.

These questions are usually worth 3 marks. To obtain the full 3 marks for these questions, your final answer must be correct.

There are three separate marks awarded for the stages of the working:

▶ Write down the correct equation needed to calculate the answer from the Relationship sheet. **(1 mark)**
▶ Show that the correct values are substituted into the equation. **(1 mark)**
▶ Show the final answer, including the correct unit. **(1 mark)**

Example of a standard '3-mark question':

The current in a resistor is 1·5 amperes when the potential difference across it is 7·5 volts. Calculate the resistance of the resistor. **(3 marks)**

Answer:

$V = IR$ **(1 mark: selecting correct relationship)**

$7·5 = 1·5 \times R$ **(1 mark: correct substitution)**

$R = 5·0\,\Omega$ **(1 mark: correct answer)**

Presenting numbers

Whenever a question requires a numerical calculation using data, important areas to consider when doing such calculations include the following.

Units

Make sure that you use the correct unit following a calculation in your final answer. **If the unit is wrong or missing, you will lose the final mark!**

Significant figures

When calculating a value using an equation, take care not to give too many significant figures in your **final answer**. This means that the final answer should have no more significant figures than the value with the least number of significant figures used in the question.

Top Tip!

Questions that ask for a calculation to be performed usually require a relationship to be selected from the Relationship sheet and applied to information given in the question.

Examples:

▶ 16·501 930 5 to 3 significant figures is 16·5
▶ 20 has 1 significant figure
▶ 40·0 has 3 significant figures
▶ 0·000 604 has 6 significant figures
▶ 4·30 × 10⁴ has 3 significant figures
▶ 6200 has 2 significant figures.

Scientific notation

When writing very large or very small numbers, use scientific notation to avoid writing or using strings of numbers in an answer or calculation.

Prefixes

You should be familiar with all common measurement prefixes, for example:

▶ nano (n) = ×10^{-9}, 1 nm = 1 × 10^{-9} m = 0·000 000 001 m
▶ micro (µ) = ×10^{-6}, 1 µm = 1 × 10^{-6} m = 0·000 001 m
▶ milli (m) = ×10^{-3}, 1 mm = 1 × 10^{-3} m = 0·001 m
▶ kilo (k) = ×10^{3}, 1 km = 1 × 10^{3} m = 1000 m
▶ mega (M) = ×10^{6}, 1 Mm = 1 × 10^{6} m = 1 000 000 m
▶ giga (G) = ×10^{9}, 1 Gm = 1 × 10^{9} m = 1 000 000 000 m

Top Tip!

When you are performing a calculation, you can use numbers with too many significant figures in your calculator. However, make sure when you give your final answer that you round your answer to match the smallest number of significant figures that appear in the data given in the question.

Top Tip!

Make sure you are familiar with how to enter and use numbers in scientific notation on your calculator before you sit your exam.

	MARKS	STUDENT MARGIN

1 Place the following quantities in the correct column of the table below.

displacement, mass, force, speed, velocity, distance, energy, acceleration, time

MARKS: 1 — STUDENT MARGIN: Dynamics 1

Scalar	Vector

2 Two forces act at right angles as shown below.

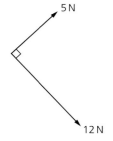

5 N

12 N

$a = \sqrt{(b^2 + c^2)}$

$a = 13$

Calculate the magnitude of the resultant force.

Space for working and answer

MARKS: 2 — STUDENT MARGIN: Dynamics 1

| | MARKS | STUDENT MARGIN |

3 An athlete runs once around the following track.

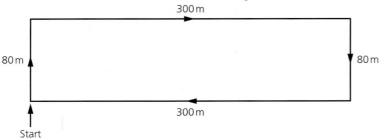

300 m

80 m   80 m

300 m

Start

Calculate

a) the total distance travelled by the athlete;

Space for working and answer

760m

1   Dynamics 1

b) the final displacement of the athlete.

Space for working and answer

0 m

1   Dynamics 1

4 A skydiver jumps from an aircraft. After some time, the parachute is opened.
The graph shows the motion of the skydiver from leaving the aircraft until landing.

2   Dynamics 2

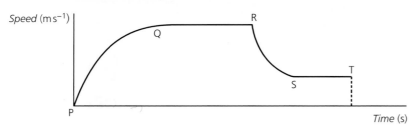

Speed (m s⁻¹)

Q   R

P   S   T

Time (s)

Which parts of the graph show when the forces acting on the skydiver are unbalanced?

Space for working and answer

PQ & RS

5 The following graph represents the motion of a speedboat travelling across a bay.

3   Dynamics 2

Speed (m s⁻¹)

20
16
12
8
4
0
0   500   1000
Time (s)

Calculate the total distance travelled by the boat.

Space for answer

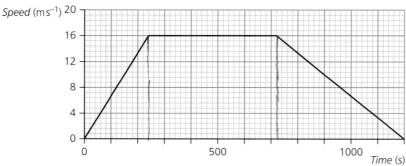

$a = \frac{1}{2}bh$
$= 0.5 \times 250 \times 16$
$a = 2000$

$a = bh$
$= 500 \times 16$
$a = 8000$

$a = \frac{1}{2}bh$
$= 0.5 \times 480 \times 16$
$= 3840$

13840 m

6 While exploring the Moon, an astronaut dropped an object onto the Moon's surface. The object took 0·8s to fall to the surface.
Calculate the vertical speed of the object just before it struck the Moon's surface.
Space for working and answer

3 — Dynamics 3

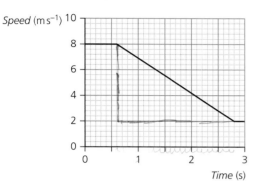

$$V = \frac{1.2}{0.8}$$
$$= 1.5$$

$$S = ut$$
$$= 1.5 \times 0.8$$
$$S = 1.2 \text{ ms}^{-1}$$

7 The speed–time graph represents the motion of a car before and after braking.

3 — Dynamics 3

Speed (m s⁻¹)

[graph with axes Speed (ms⁻¹) 0–10 vertical, Time (s) 0–3 horizontal]

Time (s)

Calculate the deceleration of the car during braking.
Space for working and answer

$$A = \tfrac{1}{2} bh$$
$$= 0.3 \times 2.2 \times 6$$
$$= 6.6$$

8 A weather balloon of total mass 40 kg is released and rises vertically from the ground.

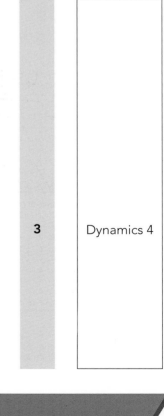

Upthrust

Weight

a) Calculate the total weight of the balloon.
Space for working and answer

3 — Dynamics 4

$$W = mg$$
$$= 40 \times 9.8$$
$$W = 392$$

	MARKS	STUDENT MARGIN

b) The upthrust exerted on the balloon when released is 510 N.
Calculate the acceleration of the balloon when released.

Space for working and answer

$510 - 392 = 118$

$F = ma$

$a = \dfrac{F}{m} = \dfrac{118}{40} = 2.95 \, ms^{-1}$

MARKS: 4 **STUDENT MARGIN:** Dynamics 4

9 A car rolls down a slope and reaches a horizontal surface.
The car has a mass 650 kg and starts at the top of a 7·2 m high slope.

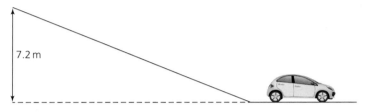

7.2 m

a) Calculate the gravitational potential energy of the car at the top of the slope.

Space for working and answer

$E_p = mgh$

$= 650 \times 9.8 \times 7.2$

$E_p = 45864 \, J$

MARKS: 3 **STUDENT MARGIN:** Dynamics 5

b) Calculate the speed of the car at the bottom of the slope, assuming no energy losses.

Space for working and answer

$\dfrac{45864 \times 0.5}{650} = 35.28 \, ms^{-1}$

MARKS: 3 **STUDENT MARGIN:** Dynamics 5

10 A baseball player strikes a ball. The ball leaves the bat horizontally at 28 m s⁻¹.
The ball hits the ground at a distance of 62 m from the point where it was struck.

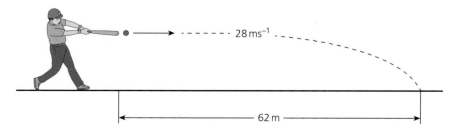

28 ms⁻¹

62 m

a) Calculate the time of flight of the ball.

Space for working and answer

$$t = \frac{d}{v}$$

$$t = \frac{60}{28}$$

$$t = 2.14 \text{ seconds}$$

b) Calculate the vertical speed of the ball as it reaches the ground.

Space for working and answer

11 Information about solar-orbiting planets is given in the table.

Planet	Distance from the Sun (×10⁹ m)	Period (days)	Mass (multiple of Earth masses)
Earth	150	365	1
Jupiter	780	4300	318
Mars	228		0·11
Mercury	58	88	0·06
Saturn	1430	10 760	95
Venus	110	225	0·82

a) Estimate a value, in days, for the period of Mars.

Space for working and answer

b) Calculate the time taken for light from the Sun to reach Jupiter.

Space for working and answer

MARKS | STUDENT MARGIN

12 Ion drive engines are used in space exploration spacecraft to provide thrust by emitting a beam of gas particles.

NASA's Dawn mission to the asteroid belt used a spacecraft with ion drive engines to propel the spacecraft while in deep space.

At full thrust, the ion engines could accelerate the spacecraft from rest to $26\,m\,s^{-1}$ in 4 days.

a) Calculate the acceleration of the spacecraft.

Space for working and answer

3 | Space 1

b) The mass of the spacecraft was 750 kg.
Calculate the size of the thrust causing the acceleration.

Space for working and answer

3 | Space 1

13 The graph shows how the gravitational field strength varies with height above the surface of the Earth.

4 | Space 1

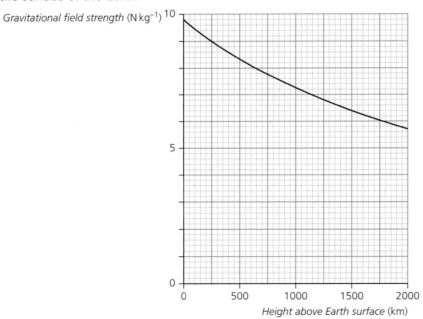

Gravitational field strength $(N\,kg^{-1})$

Height above Earth surface (km)

A satellite orbiting the Earth at an altitude of 1250 km has a weight of 306 N.

Calculate the mass of the satellite.

Space for working and answer

	MARKS	STUDENT MARGIN

14 At lift-off, a rocket has a total mass of $2{\cdot}05 \times 10^6$ kg.

The resultant force acting upwards on the rocket is $8{\cdot}2 \times 10^6$ N.

Calculate the acceleration of the rocket at lift-off.

Space for working and answer

3 — Space 1

15 The Black Eye galaxy is approximately 17 million light years from Earth.

Calculate the distance of the galaxy from Earth in metres.

Space for working and answer

3 — Space 2

16 A rechargeable battery pack stores 17 640 C of charge when fully charged. After recharging two smartphones, the battery pack is completely discharged.

The battery pack can be recharged using a charger that supplies a constant current.

The charger is used to fully recharge the battery pack in $3{\cdot}5$ hours.

Calculate the value of the constant charging current supplied by the charger.

Space for working and answer

3 — Electricity 1

17 The atomic particles neutrons, protons and electrons are directed into an electric field between two metal plates as shown.

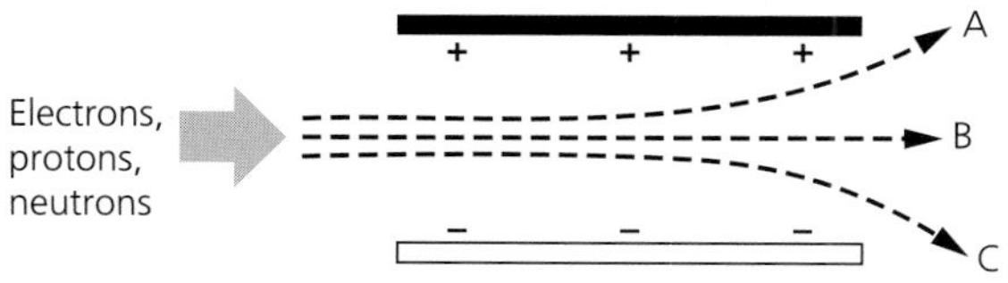

Explain which path is taken by each particle.

A _____

B _____

C _____

3 — Electricity 2

18 A circuit containing a light-dependent resistor (LDR) is shown.

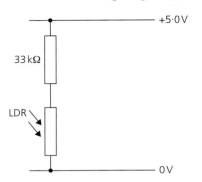

The resistance of the LDR for different conditions is shown in the table.

LDR	Resistance of LDR (kΩ)
covered	44
uncovered	4

Calculate the voltage across the LDR when it is covered.

Space for working and answer

19 An automatic washbasin tap used in a washroom is shown in the diagram.

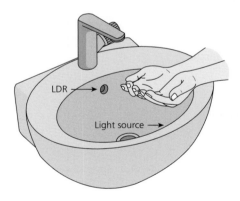

Inserting hands into the washbasin breaks a light beam; this is detected using a light-dependent resistor (LDR). The LDR is part of a switching circuit that activates the water tap valve, when hands are placed under the tap, to switch on the water.

Part of the circuit for the washbasin tap is shown below.

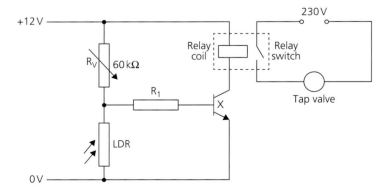

		MARKS	STUDENT MARGIN

a) Name component X in the circuit diagram.

1   Electricity 4

b) Explain how this circuit operates to activate the water tap valve when the light level falls below a certain value.

3   Electricity 4

20 Part of a circuit is shown below.

3   Electricity 4

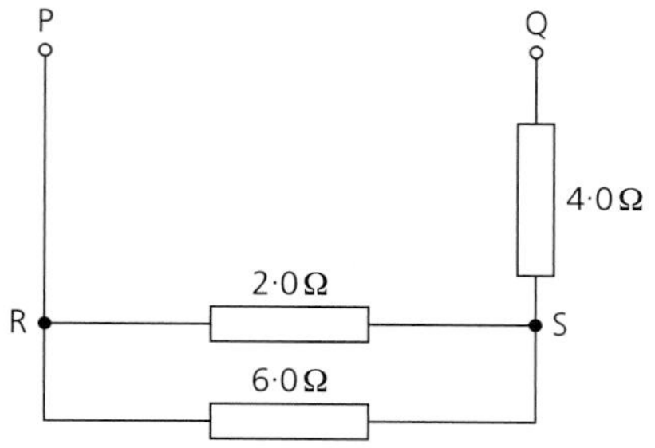

Calculate the total resistance between points P and Q.

Space for working and answer

21 A doorbell contains a buzzer. The buzzer emits a sound when the door switch is pressed.

The buzzer has a resistance of 120 Ω and a power of 147 mW.

Calculate the voltage across the buzzer when it sounds.

Space for working and answer

3   Electricity 5

22 A technician tests an electric kettle. The kettle is filled with water and switched on.

The technician records the following information:

Initial water temperature	8 °C
Final water temperature	80 °C
Total heat energy supplied to water	376 200 J

	MARKS	STUDENT MARGIN

a) Calculate the mass of water in the kettle.

Space for working and answer

MARKS: **3**

STUDENT MARGIN: Properties of matter 1

b) Explain why the mass of water is likely to be different from that calculated in part **a)**.

MARKS: **1**

STUDENT MARGIN: Dynamics 5

23 The water in a kettle reaches boiling point.
Before the kettle switches off, 420 000 J of heat energy continues to be supplied to the boiling water.
Calculate the mass of water changed into steam.

Space for working and answer

MARKS: **3**

STUDENT MARGIN: Properties of matter 2

24 A solid block of mass 50 kg has dimensions 3 m by 2 m by 0·5 m.

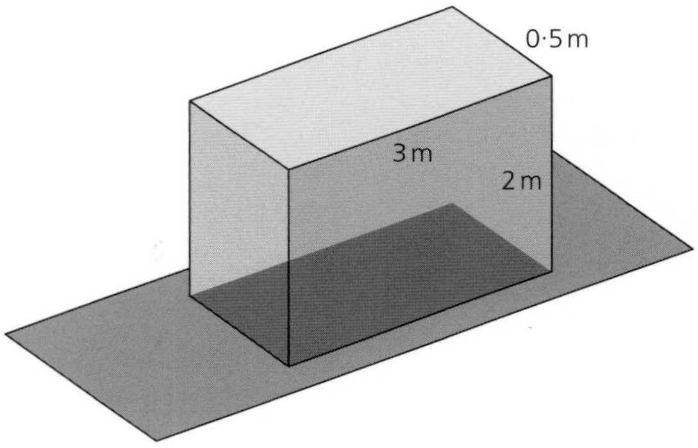

What is the smallest pressure that the block can exert on the flat horizontal surface?
Space for working and answer

MARKS: **3**

STUDENT MARGIN: Properties of matter 3

	MARKS	STUDENT MARGIN

25 A sample of gas in a sealed container is at a pressure of 180 kPa and at a temperature of 145 °C.

Calculate the new pressure if its temperature is reduced to 68 °C and the volume of the gas remains constant.

Space for working and answer

3 — Properties of matter 3

26 The following diagram gives information about a wave.

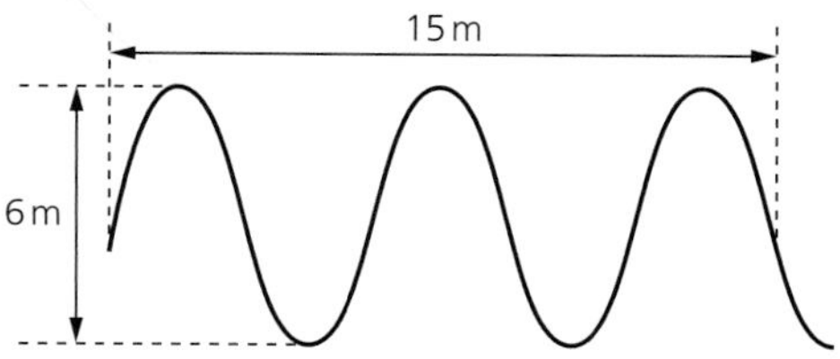

a) State the amplitude of the wave.

1 — Waves 1

b) Calculate the wavelength of the wave.

Space for working and answer

1 — Waves 1

27 A portable radio is used to send a radio signal to the emergency services.
The frequency of the signal is 2100 MHz
Calculate the wavelength of this signal.
Space for working and answer

3 — Waves 1

28 Water waves reach the entrance to a marina as shown.

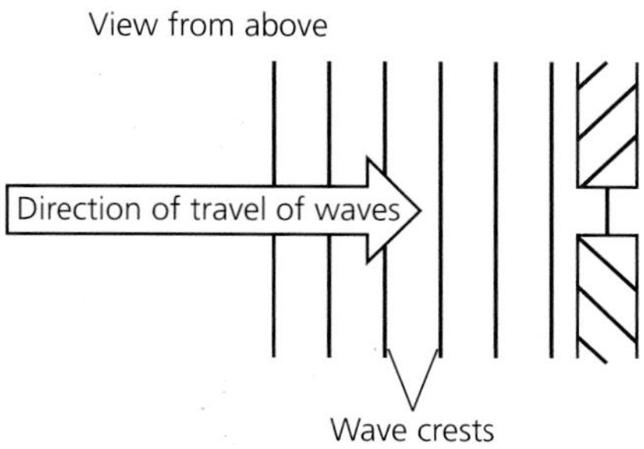

Complete the diagram to show the pattern of wave crests inside the marina.

2 — Waves 1

		MARKS	STUDENT MARGIN

29 The diagram shows the electromagnetic spectrum in order of increasing frequency.

TV and radio	P	Q	Visible light	R	X-rays	Gamma rays

Increasing frequency (Hz) →

Name the radiation:

P _____

Q _____

R _____

MARKS: 2   **STUDENT MARGIN:** Waves 2

30 a) Complete the diagram to show the path of the ray of light through the block and after it emerges from the block.

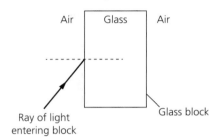

Air   Glass   Air

Ray of light entering block

Glass block

MARKS: 2   **STUDENT MARGIN:** Waves 3

b) On your diagram indicate an angle of refraction, *r*.

MARKS: 1   **STUDENT MARGIN:** Waves 3

31 A radioactive source is used to carry out checks on aircraft welding joints for cracks inside the metal frame of the aircraft.

The checks usually last around 24 hours.

The following radioactive sources are available.

Source	Half-life	Radiation emitted
W	20 years	alpha
X	15 hours	beta
Y	30 years	gamma
Z	3 hours	gamma

Explain which source would be most suitable for the purpose of detecting cracks in the aircraft's metal frame during the checks.

MARKS: 2   **STUDENT MARGIN:** Radiation 1

	MARKS	STUDENT MARGIN

32 The half-life of a radioactive source is 128 years.

In 5 hours, $4 \cdot 14 \times 10^8$ radioactive nuclei in the source decay.

Calculate the activity of the source.

Space for working and answer

3 Radiation 1

33 A worker using an X-ray machine has a mass of 90 kg and on a particular day absorbs 8·2 mJ of energy from the X-ray machine.

a) Calculate the absorbed dose received by the worker.

Space for working and answer

3 Radiation 1

b) Calculate the equivalent dose received by the worker.

Space for working and answer

3 Radiation 1

34 A radioactive source has a half-life of 12 hours.

The source has an initial activity of 128 MBq.

Calculate its activity after 2 days.

Space for working and answer

2 Radiation 1

Question type: Extended

≫ HOW TO ANSWER

In the National 5 Physics exam, Section 2 contains restricted and extended-response questions worth from 3 to around 12–16 marks each, totalling 110 marks altogether.

The marks are distributed proportionately across all six topics of the course content. In Section 2, 70–80 marks are for the demonstration and application of knowledge based on the course content. The remaining 30–40 marks are for the application of skills of scientific inquiry.

As with Section 1, 1 mark should take just over a minute (67 seconds) but questions with lots of reading, thinking time or those with calculations or information to process will take longer and other questions may be quicker.

Questions are taken from all six areas of the course. The number of marks for questions from each area is approximately in proportion to its content or size. Extended questions usually consist of several parts and require you to apply your knowledge and skills, from one or more areas.

Usually, the mark allocation and the space provided gives an indication of what length of response is required. Each individual mark is awarded separately for statements or explanations, so if a question is worth 2 marks there will be two parts required for the answer.

The questions in Section 2 usually use key phrases which indicate the type of response required. These are called 'Command terms'. The headings below show some commonly used command terms, followed by how to answer these kinds of questions.

Calculate

Use a relationship from the Relationship sheet to calculate a value for the quantity in the question.

▶ Write down the appropriate relationship as written in the Relationship sheet.
▶ Convert any number in the question to standard units (for example, km to m), preferably using scientific notation (for example, km to $\times 10^3$ m).
▶ Substitute the (converted) numbers from the question into the relationship.
▶ Use a calculator to perform the calculation.
▶ Write down the answer with units to the appropriate precision, i.e. the fewest number of significant figures in the question.

This is usually a 3-mark question, with marks given for:

▶ selection of correct relationship (1 mark)
▶ number substitution (1 mark)
▶ correct answer with units. (1 mark)

Full marks will be given for just the correct answer with units but it is always better to show the complete working in order to avoid lost marks for errors in the final answer.

Show that

Similar to calculate, but the answer (usually a numerical value with a unit) is given in the question.

The approach should be the same as for 'calculate' above – start by writing down the equation that leads to the calculation, followed by the substitution of values from the question, then showing the final answer and units. Note that this answer will most likely be used in the subsequent part of a question. The reason for a 'show that' style question is to allow candidates to gain access to the later marks in a question.

Determine

In this type of question, you may be asked to read the values for a particular point on a graph or calculate the gradient of the line.

Write down the answer with units.

Draw

Draw a graph to show the shape of the relationship between the dependent and independent variables in the question. Your graph axes must:

▶ have labels with units

▶ show the origin (as a zero) and have suitable scales – do not use a scale with a 'hidden' origin.

These are usually 3-mark questions with marks awarded for:

▶ providing suitable scales

▶ labelling the axes correctly

▶ plotting the data points and drawing the best-fit straight line or curve, as appropriate.

When asked to draw a graph, use the graph paper provided after the question. Make sure you have a sharp pencil, a ruler and a calculator to help with these questions. Try to use most of the graph paper when forming the scales. There is always a 'spare' graph paper at the end of the exam paper if you need to redraw. If used, remember to label the graph with the question number!

State (what is meant by)

Describe a definition of a physical term or phenomenon from the course specification document.

This is often the first part of a multi-part question and subsequent parts will contain related calculations. It is mainly recall and you should try to keep the definition precise by memorising the definitions in the course specification.

State (the value of)

Write down the value of the quantity with units and to appropriate precision.

The value will be read from a graph or may have already been calculated in an equivalent quantity previously in the question.

A common example might be when the first part of a question asks you to calculate the gravitational potential energy of an object. The next part might ask you to state the value of the kinetic energy of the object when it falls to the ground by equating final kinetic energy to the loss of potential energy for a falling object (in the absence of 'losses' due to friction).

Explain (what happens to …)

Typically, this is the last part of a multi-part question.

Sometimes this is asked as:

▶ 'State what happens to …'

▶ 'You must justify your answer.'

You may find an equation that has been used previously in the question can help your justification of your answer.

Top Tip!

No marks are awarded without an attempted justification for this type of question.

This usually requires you to think about the description given, then to select an appropriate answer using your knowledge of the coursework, or from a table of possible substances.

For example:

> *The following table shows which radioactive sources are available.*
>
> *State which radioactive source should be used.*
>
> *You **must** explain your answer.*

Many of the extended questions in Section 2 of the exam will contain familiar situations, but some may contain unfamiliar situations, for example, referring to new technology or a new invention. Where this is the case, you may need to interpret the information based on your knowledge of the coursework and problem solve in order to answer the question. Although the application may be new to you, the question will be answerable by applying your knowledge of the coursework and the skills developed during your studies.

Top Tip!

If a question has an unfamiliar setting, it will still be answerable by applying your coursework knowledge and skills.

	MARKS	STUDENT MARGIN

1   A cruise ship is sailing due East at a constant speed on a calm sea.

The cruise ship and its passengers have a total mass of 9.7×10^7 kg.

a)   Calculate the total weight of the cruise ship and its passengers.

Space for working and answer

MARKS: **3**

STUDENT MARGIN: Dynamics 4

b)   While sailing, the cruise ship's engines produce a force of 4.6×10^3 N due East. The cruise ship encounters a strong tide from North to South, which exerts a force of 1.8×10^3 N.

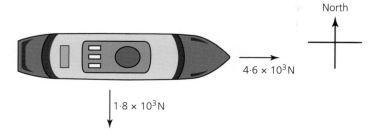

North

4.6×10^3 N

1.8×10^3 N

By scale diagram, or otherwise, determine:

(i) the magnitude of the resultant force acting on the cruise ship;   2   Dynamics 1
    Space for working and answer

(ii) the direction of the resultant force acting on the cruise ship.   2   Dynamics 1
    Space for working and answer

2 A cyclist attempted to beat a record time for cycling along a straight level cycle track. The attempt was recorded by a news camera team on a motorcycle alongside the cyclist.

The graph shows the speed of the cyclist during the attempt.

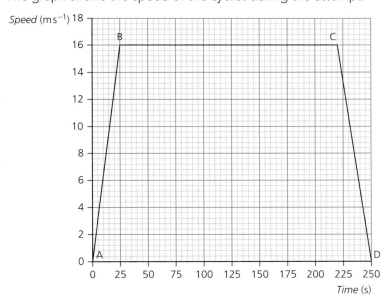

a) **(i)** Calculate the acceleration of the cyclist between C and D.   3   Dynamics 3
    Space for working and answer

	MARKS	STUDENT MARGIN

(ii) Describe the motion of the cyclist between B and C.
1 — Dynamics 2

(iii) On the drawing of the cyclist, draw and name the horizontal forces acting on the cyclist between B and C.
You **must** indicate the direction of each force.
2 — Dynamics 4

b) **(i)** Calculate the total distance travelled by the cyclist.
Space for working and answer
3 — Dynamics 2

(ii) Calculate the average speed of the cyclist.
Space for working and answer
3 — Dynamics 1

3 A student demonstrates an experiment to illustrate projectile motion.

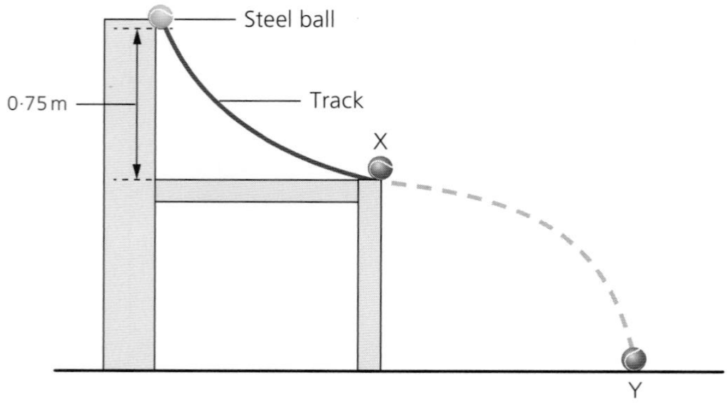

Steel ball
0·75 m — Track
X
Y

A steel ball of mass 0·20 kg is released from the top of a track. The ball rolls down the track then leaves the track horizontally at position X and lands on the ground at position Y.

	MARKS	STUDENT MARGIN

a) The ball takes 0·55 s to reach the ground after leaving position X. Calculate the vertical velocity of the ball as it reaches the ground. Ignore the effect of air resistance.

Space for working and answer

3   Dynamics 6

b) Determine the height of position X above the ground.

Space for working and answer

4   Dynamics 6

c) In a second experiment, the student drops the ball vertically from position X to the ground.

1   Dynamics 6

State how the time taken for the ball to reach the ground from X compares with the time taken for the ball in part **a)**.

4 a) The table gives information about some of the planets in our solar system.

	Planet						
	Neptune	Saturn	Jupiter	Mars	Earth	Venus	Mercury
Time to orbit the Sun once (in Earth years)	165	29	12	1·9	1	0·6	0·24
Time for one complete spin (in Earth days or Earth hours)	16 hours	10 hours	10 hours	25 hours	24 hours	243 days	59 days
Gravitational field strength (N kg⁻¹)	11	9·0	23	3·7	9·8	8·9	3·7
Distance from the Sun (million km)	4500	1430	780	228	150	110	58

(i) On which planets will a 5 kg mass have the same weight?

1   Space 1

(ii) Which planets have the same time for one day?

1   Space 1

(iii) Which planet takes longest to orbit the Sun?

1   Space 1

b) Exploration of other planets requires vast distances to be travelled. Once launched, a spacecraft has to reach a high velocity to reach a distant destination in a reasonable time. The spacecraft itself has limited capacity for carrying conventional fuel.

(i) Explain how an ion drive engine might be the possible solution for a spacecraft to attain a high velocity in space.

2   Space 1

(ii) Explain how a 'gravity assist' or 'catapult' from a moon or planet could increase the velocity of a passing spacecraft.

2   Space 1

c) Maintaining sufficient energy to operate life support systems in a manned spacecraft mission is another challenge of space travel.
Suggest a possible solution to this challenge.

1

5 The brightness of three identical lamps is controlled in the circuit shown.
The variable resistor is adjusted until the lamps operate at their correct voltage of 4·5V.

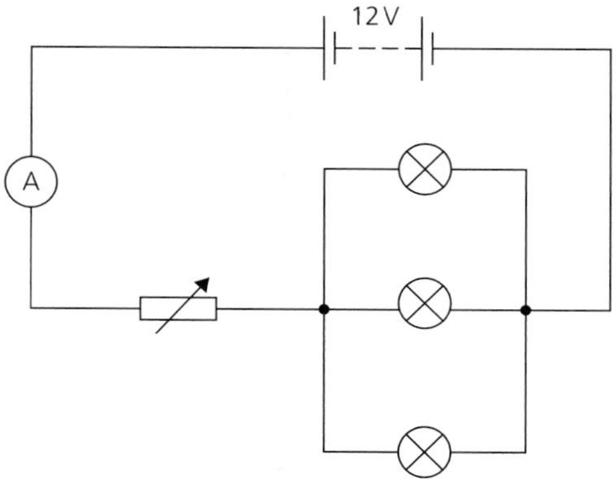

a) The reading on the ammeter is 1·5A when the lamps operate at the correct voltage.
Calculate the current in one lamp.

Space for working and answer

1

Electricity 4

b) Calculate the power developed in one lamp when operating at the correct voltage.

Space for working and answer

3

Electricity 5

c) An identical lamp is added in parallel with the three lamps.
What happens to the reading on the ammeter?
You must justify your answer.

2

Electricity 4

MARKS   STUDENT MARGIN

6 A light-dependent resistor (LDR) is used as a light sensor in a circuit to monitor the light level outside a greenhouse. When the light level outside the greenhouse falls below a certain level, lamps are switched on inside the greenhouse. Part of the circuit is shown.

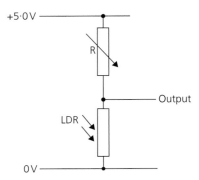

a) **(i)** The variable resistor R is set at a resistance of 2250 Ω.

Calculate the resistance of the LDR when the voltage across the LDR is 2·0 V.

Space for working and answer

4   Electricity 3

(ii) The graph shows how the resistance of the LDR varies with the outside light level.

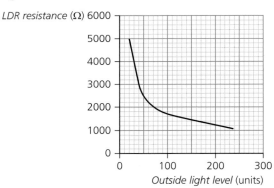

Use the graph to determine the outside light level when the voltage across the LDR is 2·0 V.

1

38

	MARKS	STUDENT MARGIN

b) The circuit is now connected to a switching circuit to operate the lamps inside the greenhouse.

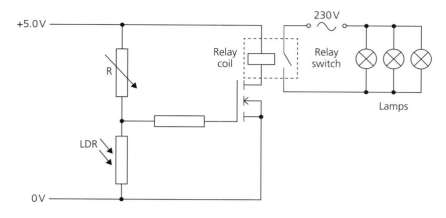

(i) Explain how the circuit operates to switch on the lamps when the outside light level falls below a certain value.

3 | Electricity 4

(ii) The resistance of the variable resistor R is now increased.
What effect does this have on the outside light level the lamps switch on at?

You must justify your answer.

3 | Electricity 4

7 A deep fat fryer is used in a kitchen to fry vegetables.
The rating plate of the deep fat fryer is shown.

Electricity 4

frequency 50 Hz

voltage 230 V

power 1500 W

a) The deep fat fryer contains 2·8 kg of vegetable oil at an initial temperature of 20 °C. The specific heat capacity of the oil is 1800 J kg^{-1} °C^{-1}.
Calculate the energy required to raise the temperature of the oil to 170 °C.

Space for working and answer

3 | Properties of matter 1

	MARKS	STUDENT MARGIN

b) Calculate the minimum time required to heat the oil to 170 °C.

Space for working and answer

3 Electricity 5

c) In practice it requires more time than calculated to heat the oil.

   (i) Explain why more time is required.

1 Dynamics 5

   (ii) Suggest one way of reducing this additional time.

1 Dynamics 5

8 A student investigates the relationship between the pressure and temperature of a fixed mass of gas at constant volume. A diagram of the apparatus used for the investigation is shown.

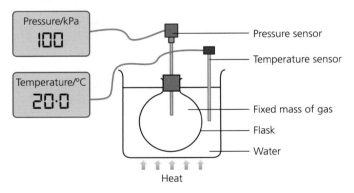

The gas temperature and pressure are obtained using sensors connected to displays.

The gas is slowly heated and the displayed results are recorded in a table.

The volume of the gas remains constant during the investigation.

Some of the results are shown.

Pressure (kPa)	100			
Temperature (°C)	20·0	28·0		

a) Calculate the pressure of the gas when the temperature is 28·0 °C.

Space for working and answer

3 Properties of matter 3

b) Use the kinetic model to explain the change in pressure as the temperature increases.

3   Properties of matter 3

c) Suggest a change that can be made to the apparatus to improve the experiment.

1   Properties of matter 3

9 Refraction of light occurs in spectacle glass lenses.

a) Explain what is meant by the term *refraction*.

1   Waves 3

b) The diagram shows a light ray entering a glass block.

(i) Complete the diagram to show the path of the light ray inside the block and after it emerges from the block.

2   Waves 3

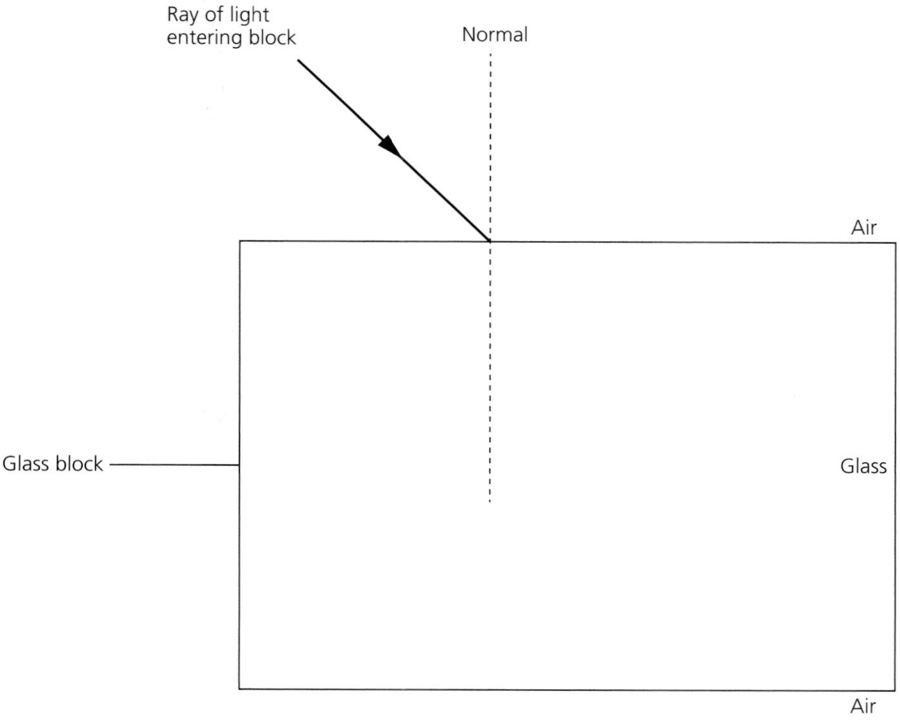

Ray of light entering block

Normal

Air

Glass block

Glass

Air

(ii) Indicate an angle of incidence, *i*, on the diagram.

1   Waves 3

MARKS   STUDENT MARGIN

10 The electromagnetic spectrum is shown in order of increasing frequency.

Television and radio waves	Microwaves	X	Visible light	Y	X rays	Gamma rays

Increasing frequency →

a) Identify radiations X and Y.

X _____

Y _____

2   Waves 2

b) Radio waves are sent from Earth to Jupiter. Jupiter is $6·7 \times 10^{11}$ m from Earth. Calculate the time taken for radio waves from Earth to reach Jupiter.

Space for working and answer

3   Waves 1

11 A news reporter in a remote area uses a satellite phone that uses microwaves to communicate via satellite with the news head office.

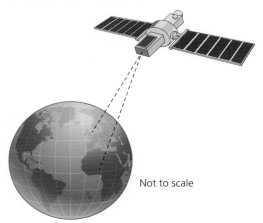

Not to scale

The period of the microwaves is $4·0 \times 10^{-10}$ s.

a) Calculate the frequency of the microwaves.

Space for working and answer

3   Waves 1

b) Calculate the wavelength of the microwaves.

Space for working and answer

3   Waves 1

MARKS   STUDENT MARGIN

12 A university technician is investigating a radioactive material.

The activity of the material over a period of time is shown in the graph.

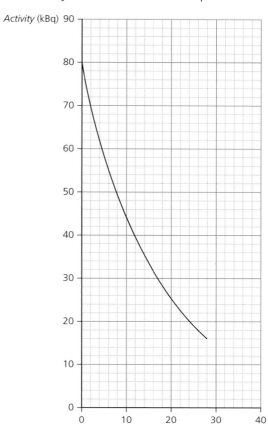

a)   **(i)** State what is meant by the term *half-life*.   **1**   Radiation 1

(ii) Use the graph to determine the half-life of the radioactive material.   **1**   Radiation 1
Space for working and answer

(iii) The initial activity of the material is 80 kBq.   **2**   Radiation 1
Determine the activity of the material after three half-lives.
Space for working and answer

b) While working with the radioactive material for 3 hours, the technician's hands receive an absorbed dose 15·0 µGy. The radiation from the material has a radiation weighting factor of 2.

(i) Calculate the equivalent dose received by the technician's hands.

Space for working and answer

(ii) Calculate the equivalent dose rate received by the technician's hands.

Space for working and answer

Question type: Experimental and data-handling

>> HOW TO ANSWER

This type of question requires you to apply your knowledge and skills to a practical or experimental situation. Usually, there is a brief explanation of an experiment with a diagram, which may be followed by a table of results and/or a graph.

In an experimental and/or data-handling question, you may be asked to:

▶ take information from the table or graph and to calculate answers using a related equation.

▶ draw a graph. When asked to draw a graph, use the graph paper provided after the question. Make sure you have a sharp pencil, a ruler and a calculator to help with these questions. This type of question is usually worth 3 marks.

▶ predict results for situations not tested, for example, carrying out the experiment at a different height, speed or pressure. Predictions can involve extending graph lines or reading between values in a table. Sometimes, the final part of the question asks for a suggested improvement to the experiment.

Top Tip!

If you are given space for a calculation, you will very likely need to use it!

For all calculations make sure that you:

▶ include units

▶ give values to the appropriate precision.

Top Tip!

In graph-drawing questions, marks are given for:

▶ providing suitable scales

▶ labelling the axes correctly

▶ plotting the data points and drawing the best-fit straight line or curve.

Top Tip!

Draw graphs using a ruler and use the data table headings and units for the axis labels. Straight-line graphs usually require a 'best-fit' straight line to be drawn using a ruler.

MARKS **STUDENT MARGIN**

1 A trolley is released from rest near the top of a track and moves downward. A card attached to the trolley passes through a light gate near the bottom of the track.

The following information is recorded.

Length of the card = 50 mm

Distance travelled by the trolley down the track = 1·4 m

Time for the card to pass through the light gate = 0·085 s

MARKS | STUDENT MARGIN

a) Calculate the instantaneous speed of the trolley as it passes through the light gate.

Space for working and answer

3 | Dynamics 1

b) Suggest a possible value for the average speed of the trolley over the 1·4 m distance travelled by the trolley down the track.

Space for working and answer

1 | Dynamics 1

2 Two identical hot water tanks are fitted with the same thickness of different types of insulation.

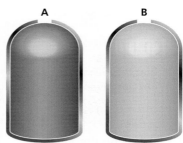

Insulated hot water tanks

Both tanks are fitted with a 4 kW water heater and filled with 80 kg of water.

Each tank has a thermostat attached to it. The thermostat switches on the heater when the temperature of the water falls to 40 °C and switches off the heater when the temperature rises to 80 °C.

The temperature–time graphs, A and B, show **one** heating cycle for each tank.

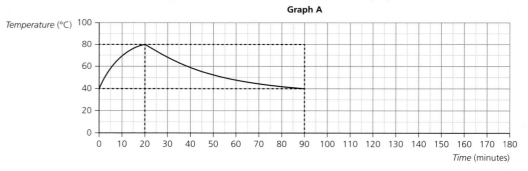

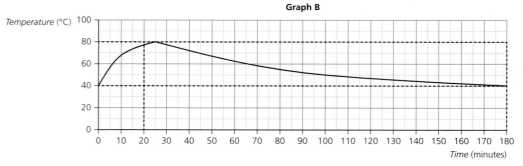

	MARKS	STUDENT MARGIN

Which tank has the best insulation?

You must justify your answer.

3 A space research team construct a small test space rocket to investigate a particular design.

The rocket contains an engine and fuel.

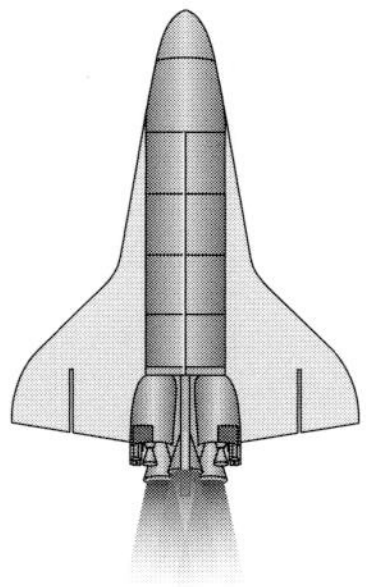

At launch, the mass of the rocket and fuel is 0·4 kg.

a) **(i)** The initial acceleration of the rocket is 45 m s⁻².

Calculate the unbalanced force acting on the rocket at this time.

Space for working and answer

3   Space 1

(ii) The rocket engines supply a constant force during the ascent.

Explain whether the acceleration of the rocket increases, decreases or remains constant.

2   Space 1

b) The table shows information about four different satellites.

Satellite	Altitude (km)	Period (minutes)
Aqua	705	99
NAVSTAR 53	20 000	
Galileo	23 222	847
Echostar 19	35 788	1440

Predict the period of the NAVSTAR 53 satellite.

2   Space 1

4   A student constructs an electrical circuit that includes resistors.

Resistor values and power ratings are indicated on the resistor by a series of coloured bands.

The student selects a resistor with a resistance labelled '20 Ω, 5 W'.v

The student decides to check the value of the resistance.

a)   Draw a circuit diagram, including a 12 V battery, a voltmeter and an ammeter, for a circuit that could be used to determine the resistance.

Space for drawing

         2   Electricity 3

b)   Readings from the circuit give the voltage across the resistor as 12·0 V and the current through the resistor as 0·60 A.

During this experiment, the resistor becomes very hot and gives off smoke. Explain why this happens.

You **must** include a calculation as part of your answer.

Space for working and answer

         3   Electricity 5

5   A student sets up the following circuit to investigate the resistance of resistor R.

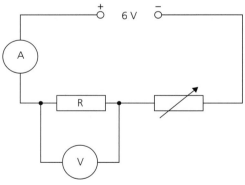

The variable resistor is adjusted and the voltmeter and ammeter readings are noted. The following graph is obtained from the experimental results.

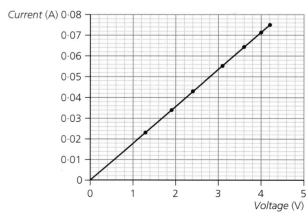

	MARKS	STUDENT MARGIN

a) (i) Calculate the value of the resistor R when the reading on the voltmeter is 4·2V.

Space for working and answer

4   Electricity 3

(ii) Using information from the graph, state whether the resistance of the resistor R increases, stays the same or decreases as the voltage increases.

Justify your answer.

2   Electricity 3

b) The student is given a task to combine two resistors from a pack containing one each of 33Ω, 56Ω, 82Ω, 150Ω, 270Ω and 390Ω.
Show by calculation which two resistors should be used to give the smallest combined resistance.

Space for working and answer

3   Electricity 4

MARKS   STUDENT MARGIN

6   A student is training to become a diver.

The student carries out an experiment to investigate the relationship between the pressure and volume of a fixed mass of gas using the apparatus shown.

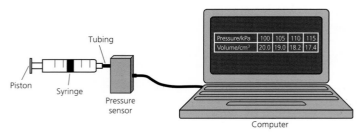

The pressure of the gas is recorded using a pressure sensor connected to a computer. The volume of the gas is also recorded. The student pushes the piston to alter the volume and a series of readings is taken.

The temperature of the gas is constant during the experiment.

The results are shown.

Pressure (kPa)	100	105	110	115
Volume (cm³)	20·0	19·0	18·2	17·4

a)   Using **all** the data, establish the relationship between the pressure and volume of the gas.

Space for working and answer

2

Properties of matter 3

b)   Use the kinetic model to explain the change in pressure as the volume of gas decreases.

3

Properties of matter 3

c)   Explain why it is important for the tubing to be as short as possible.

1

Properties of matter 3

		MARKS	STUDENT MARGIN

7 **a)** Complete the table below, by correctly inserting the following **three** types of wave.
light wave, sound wave, radio wave

2 — Waves 1

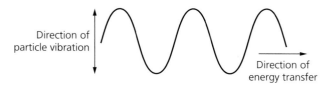

Transverse	Longitudinal

b) The diagram below represents a waveform.

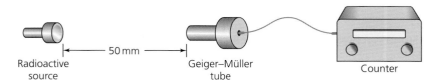

Direction of particle vibration

Direction of energy transfer

(i) Explain which type of wave is represented.

1 — Waves 1

(ii) The wavelength of these waves is 2 m. Ten waves are produced every 5 seconds.
Calculate the speed of the waves.
Space for working and answer

4 — Waves 1

8 A technician sets up an experiment as shown to identify the type of radiation being emitted from a radioactive source.

Radioactive source

50 mm

Geiger–Müller tube

Counter

The technician places a 4 mm sheet of aluminium between the radioactive source and the Geiger–Müller tube. The count rate is observed to decrease and the technician concludes that the radioactive source is emitting beta radiation.

a) Suggest **one** reason why the technician's conclusion may be incorrect.

1 — Radiation 1

b) Suggest how the experiment might be improved to identify the radiation being emitted.

1 — Radiation 1

c) State **two** safety precautions that the technician should observe when handling radioactive sources.

1 — Radiation 1

Question type: Open-ended

>> HOW TO ANSWER

There is a maximum of two open-ended questions in Section 2. They are usually stand-alone questions, but sometimes they form a part of a more extended question. They are worth a maximum of 3 marks each and the marks awarded depend on the depth of your answer.

The open-ended question usually discusses a physics phenomenon and usually asks you to '**use your knowledge**' of physics to explain it. You have to think about the issue and give a step-by-step logical answer. There may be more than one area of physics used to answer this type of question. There can be a number of acceptable answers for this type of question.

When you answer an open-ended question:

▶ Try to make three relevant comments about the context of the question – these can be bullet points. Your answer does not have to consider every single part of physics which may apply to the description. However, you should not state anything that is wrong in terms of physics.

▶ If there is an obvious equation or relationship, write it down and/or sketch the graph as part of one of your comments.

▶ If a graph is relevant, you could also describe the effect of changing the independent variable on the dependent variable within the question context.

Be careful not to spend longer than necessary on these 3-mark questions – up to 5 minutes is a good guide.

Top Tip!

The number of marks awarded will depend on how much your answer demonstrates your understanding of the physics in the question:

▶ no understanding – 0 marks

▶ limited understanding – 1 mark

▶ reasonable understanding – 2 marks

▶ good understanding – 3 marks.

	MARKS	STUDENT MARGIN
1 Car designers are constantly trying to reduce the environmental impact of cars. One way to do this is to make them more fuel efficient, as the less fuel cars need, the fewer dangerous gases they emit into the atmosphere.  Use your knowledge of physics to comment on how car manufacturers might produce cars that are more fuel efficient.	3	Dynamics 4

		MARKS	STUDENT MARGIN

2 What affects how long it takes objects to fall to the ground?

Use your knowledge of physics to answer this question.

3 | Dynamics 4

3 When a spacecraft is launched into space it accelerates to reach speeds of up to $8\,km\,s^{-1}$ to achieve orbit.

At launch, most of its mass consists of the fuel required to provide upthrust for this acceleration.

During the launch, the acceleration of the spacecraft is not constant.

Use your knowledge of physics to comment on why the acceleration is not constant.

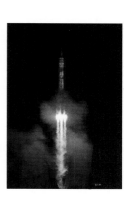

3 | Space 1

4   Estimate the pressure exerted on the floor by an average National 5 student who is standing on two feet.

Show any working clearly and explain any assumptions that you make.

3   Properties of matter 3

5   A mass of copper heated with a Bunsen is immersed in a beaker of cold water.

3   Properties of matter 1

Use your knowledge of physics to comment on what the final temperature of the copper and water would depend on.

Make reference to any relevant equation(s) in your answer.

MARKS | STUDENT MARGIN

6 Scientists collect data from different parts of the electromagnetic spectrum to obtain information about astronomical objects.

3 | Waves 2

 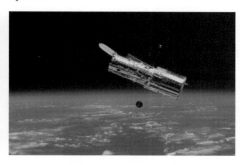

Use your knowledge of physics to comment on why different telescopes are required to obtain information about different parts of the electromagnetic spectrum.

Question type: Scientific literacy

>> HOW TO ANSWER

These are questions testing your scientific reading skills, where you will be asked about a scientific report or passage. There is usually one of this type of question in Section 2. You may already know about some of the information, but some or all of it may be new to you.

In a scientific literacy question, you may be asked to:

▶ find and write down a sentence, word or value from the passage

▶ interpret the passage, or make a prediction based on what you have read

▶ perform a calculation using data and an unfamiliar equation given in the passage. The equation should be used in exactly the same way as routine equations in the coursework.

Top Tip!

Read the passage carefully and spend around 1½ minutes per mark on this type of question.

Scientific literacy passages or reports are often new, unfamiliar situations. Don't panic! All the questions can be answered by:

▶ finding the information in the passage

▶ finding the right equations for a calculation on the Relationship sheet at the start of your examination paper

▶ applying your knowledge from the course.

MARKS   STUDENT MARGIN

1   Read the passage and answer the questions that follow.

Neutron stars

When stars reach the end of their life, they can become neutron stars. A neutron star has a mass ranging from 1·4 to 3·2 times that of our Sun. In a neutron star, this huge mass is contained within a diameter of approximately 12 km and means that a neutron star is extremely dense. One cupful of this mass would have the same weight as Mount Everest on Earth! The extreme density of the neutron star also means that it has very large gravitational and magnetic field strengths.

	MARKS	STUDENT MARGIN

Stars emit huge amounts of energy. This energy is the result of nuclear fusion happening at the centre of the star. Nuclear fusion of the isotopes of hydrogen produces helium and also the energy which sustains the star's massive shape.

Neutron stars are thought to be formed when large stars collapse. This happens when the fusion process stops and there is no longer enough energy to sustain the star. The star explodes. This explosion is known as a supernova. The outer gases of the star expand rapidly to produce an extremely bright object in the sky, which can be seen by astronomers on Earth.

The gravitational field causes the centre of the star to collapse. Its volume reduces dramatically. During the collapsing process, electrons and protons combine to form neutrons. This is the reason for the name 'neutron' stars.

Neutron stars sometimes appear in binary systems, where they are in mutual orbit around another object. X-ray telescopes on satellites have been used by astronomers to obtain data from such binary systems. This data has confirmed the mass of the neutron star to be 1·4–3·2 times that of the Sun's mass.

Neutron stars rotate rapidly when newly formed and gradually slow over a long period of time. A neutron star known as PSR J1748-2446ad rotates 716 times per second.

Some neutron stars emit radio waves or X-rays. These emissions only occur at the magnetic poles of the neutron star. When observed by astronomers, these emissions appear as 'pulses' of radio waves or X-rays. The pulses appear at the same rate as the rotation of the neutron star. Such neutron stars are known as 'pulsars'.

a) What event is thought to lead to the formation of neutron stars?

1 — Space 2

b) Why does the neutron star consist mainly of neutrons?

1 — Space 2

c) Calculate the period of rotation of the neutron star known as PSR J1748-2446ad.

Space for working and answer

3 — Waves 1

MARKS **STUDENT MARGIN**

2   Read the passage below and answer the questions that follow.

Asteroids

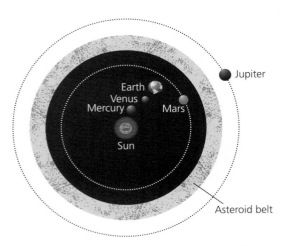

An asteroid is an object in space which orbits the Sun. There are large numbers of asteroids, which range in size from small pebbles to massive rocky objects, including the dwarf planet Ceres.

It is thought that the solar system was formed about 4·5 billion years ago, when vast clouds of interstellar dust and gas collapsed and condensed to form the Sun and planets. However, some of the remaining matter formed smaller objects which did not become part of the Sun or a planet.

Many asteroids exist in a region known as the asteroid belt. This contains millions of solar-orbiting objects. The asteroid belt is located between the orbits of planets Mars and Jupiter, around $4·2 \times 10^{-5}$ light years from the Sun.

Sometimes, gravitational effects from the larger planets – Jupiter, Saturn, Neptune or Uranus – can cause some asteroids to alter their orbit into a path that brings them close to the Earth. When these asteroids reach the Earth, they are known as meteors and burn up on entry into the Earth's atmosphere.

Some scientists believe that the craters of the Moon were formed by the strikes of asteroids.

Although there are huge numbers of asteroids, they do not pose a problem for passing space probes because of the extremely large spaces between them in the asteroid belt.

a)   When is the asteroid belt thought to have been formed?   **1**   Space 1

b)   What sometimes causes an asteroid to leave its orbit in the asteroid belt?   **1**   Space 1

c)   What is the average distance, in metres, of the asteroid belt from the Sun?   **3**   Space 2
     Space for working and answer

Multiple-choice questions

Question	Answer	Max. mark
1	B	1
2	A	1
3	C	1
4	B	1
5	B	1
6	B	1
7	D	1
8	E	1
9	B	1
10	D	1
11	B	1
12	C	1
13	B	1
14	D	1
15	C	1
16	C	1
17	E	1
18	D	1
19	E	1
20	B	1
21	D	1
22	B	1

Question	Answer	Max. mark
23	B	1
24	C	1
25	B	1
26	A	1
27	E	1
28	D	1
29	C	1
30	B	1
31	A	1
32	B	1
33	C	1
34	C	1
35	E	1
36	B	1
37	C	1
38	D	1
39	A	1
40	D	1
41	B	1
42	A	1
43	E	1
44	E	1

Short questions

Question		Answer		Max. mark
1		<table><tr><td>*Scalar*</td><td>*Vector*</td></tr><tr><td>mass speed distance energy time</td><td>displacement force velocity acceleration</td></tr></table>		1
2		$R = \sqrt{5^2 + 12^2} = 13\,\text{N}$		2
3	**a)**	$d = 80 + 300 + 80 + 300 = 760\,\text{m}$		1
	b)	Final displacement = 0		1
4		P–Q	(1)	2
		R–S	(1)	

Question		Answer		Max. mark
5		Distance = area under speed–time graph	(1)	3
		$d = \left(\frac{1}{2} \times 240 \times 16\right) + (480 \times 16) + \left(\frac{1}{2} \times 480 \times 16\right)$	(1)	
		$= 13440\,\text{m} = 13000\,\text{m}$	(1)	
6		$a = \dfrac{v - u}{t}$	(1)	3
		$1 \cdot 6 = \dfrac{v - 0}{0 \cdot 8}$	(1)	
		$v = 1 \cdot 28\,\text{ms}^{-1} = 1 \cdot 28 = 1 \cdot 3\,\text{ms}^{-1}$	(1)	
7		$a = \dfrac{v - u}{t}$	(1)	3
		$a = \dfrac{2 - 8}{2 \cdot 2}$	(1)	
		$a = -2 \cdot 7\,\text{ms}^{-2}$	(1)	
		which is a deceleration of $2 \cdot 7\,\text{ms}^{-2}$		
8	a)	$W = mg$	(1)	3
		$= 40 \times 9 \cdot 8$	(1)	
		$= 392\,\text{N} = 390\,\text{N}$	(1)	
	b)	$F_{\text{un}} = 510 - 392 = 118\,\text{N}$	(1)	4
		$F = ma$	(1)	
		$118 = 40 \times a$	(1)	
		$a = 2 \cdot 95\,\text{ms}^{-2} = 3 \cdot 0\,\text{ms}^{-2}$	(1)	
9	a)	$E_p = mgh$	(1)	3
		$= 650 \times 9 \cdot 8 \times 7 \cdot 2$	(1)	
		$= 45\,864\,\text{J} = 46\,000\,\text{J}$	(1)	
	b)	$E_k \equiv E_p$		3
		$45\,864 = \dfrac{1}{2} mv^2$	(1)	
		$45\,864 = \dfrac{1}{2} \times 650 \times v^2$	(1)	
		$v = 11 \cdot 9\,\text{ms}^{-1} = 12\,\text{ms}^{-1}$	(1)	
10	a)	$v = \dfrac{d}{t}$	(1)	3
		$28 = \dfrac{62}{t}$	(1)	
		$t = 2 \cdot 2\,\text{s}$	(1)	
	b)	$a = \dfrac{v - u}{t}$	(1)	3
		$9 \cdot 8 = \dfrac{v - 0}{2 \cdot 2}$	(1)	
		$t = 21 \cdot 6\,\text{ms}^{-1} = 22\,\text{ms}^{-1}$	(1)	
11	a)	Any number of days greater than 365 and less than 780 (you must state a number).		1
	b)	$v = \dfrac{d}{t}$	(1)	3
		$3 \times 10^8 = \dfrac{780 \times 10^9}{t}$	(1)	
		$t = 2600\,\text{s}$	(1)	

Question		Answer		Max. mark
12	a)	$a = \dfrac{v - u}{t}$	(1)	3
		$a = \dfrac{26 - 0}{4 \times 24 \times 60 \times 60}$	(1)	
		$a = 7 \cdot 5 \times 10^{-5}\,ms^{-2}$	(1)	
	b)	$F = ma$	(1)	3
		$= 750 \times 7 \cdot 5 \times 10^{-5}$	(1)	
		$= 0 \cdot 056\,N$	(1)	
13		From graph, at 1250 km altitude, the gravitational field strength is $6 \cdot 8\,N\,kg^{-1}$	(1)	4
		$W = mg$	(1)	
		$306 = m \times 6 \cdot 8$	(1)	
		$m = 45\,kg$	(1)	
14		$F_R = ma$	(1)	3
		$8 \cdot 2 \times 10^6 = 2 \cdot 05 \times 10^6 \times a$	(1)	
		$a = 4 \cdot 0\,ms^{-2}$	(1)	
15		$d = vt$	(1)	3
		$d = 3 \times 10^8 \times 17 \times 10^6 \times 365 \cdot 25 \times 24 \times 60 \times 60$	(1)	
		$d = 1 \cdot 6 \times 10^{23}\,m$		
		(Note that 365·25 days (the average number of days in one year.)	(1)	
16		$Q = It$	(1)	3
		$17\,640 = I \times 3 \cdot 5 \times 60 \times 60$	(1)	
		$I = 1 \cdot 4\,A$	(1)	
17		A Electrons are negatively charged and experience a force upwards	(1)	3
		B Neutrons have no charge and experience no force	(1)	
		C Protons are positively charged and experience a force downwards	(1)	
18		R_{LDR} (covered) $= 44\,k\Omega$	(1)	4
		$V_2 = \left(\dfrac{R_{LDR}}{R_1 + R_{LDR}} \right) Vs$	(1)	
		$V_2 = \left(\dfrac{44 \times 10^3}{33 \times 10^3 + 44 \times 10^3} \right) \times 5$	(1)	
		$V_2 = 2 \cdot 9V$	(1)	
19	a)	Transistor		1
	b)	As R_{LDR} increases, V_{LDR} increases.	(1)	3
		When V_{LDR} reaches switching voltage, the transistor turns on.	(1)	
		Current is now in the relay which completes the water tap valve circuit.	(1)	
20		Resistance between R and S:		3
		$\dfrac{1}{R_{R-S}} = \dfrac{1}{R_1} + \dfrac{1}{R_2}$	(1)	
		$= \dfrac{1}{2} + \dfrac{1}{6}$		
		$R_{R-S} = 1 \cdot 5\Omega$	(1)	
		So $R_{total} = 1 \cdot 5 + 4 = 5 \cdot 5\Omega$	(1)	

Question		Answer	Max. mark
21		$P = \dfrac{V^2}{R}$ (1) $147 \times 10^{-3} = \dfrac{V^2}{120}$ (1) $V = 4 \cdot 2\,V$ (1)	3
22	a)	$E_h = cm\Delta T$ (1) $376\,200 = 4180 \times m \times (80 - 8)$ (1) $m = 1 \cdot 25\,\text{kg}$ (1)	3
	b)	Some of the heat energy supplied to the water may be used to heat the container and some may be lost to the surroundings.	1
23		$E_h = ml$ (1) $420\,000 = m \times 22 \cdot 6 \times 10^5$ (1) $m = 0 \cdot 19\,\text{kg}$ (1)	3
24		Smallest pressure with largest area in contact with surface. $p = \dfrac{F}{A}$ (1) $p = \dfrac{490}{3 \times 2}$ (1) $p = 81 \cdot 7\,\text{Pa} = 82\,\text{Pa}$ (1)	3
25		$T_1 = 145\,°C = 145 + 273 = 418\,K$, $T_2 = 68\,°C = 68 + 273 = 341\,K$ $\dfrac{P_1}{T_1} = \dfrac{P_2}{T_2}$ (1) $p_2 = \dfrac{180 \times 341}{418}$ (1) $= 146 \cdot 8\,\text{kPa} = 147\,\text{kPa}$ (1)	3
26	a)	3 m	1
	b)	$\lambda = \dfrac{\text{length of waves}}{\text{number of waves}} = \dfrac{15}{2 \cdot 5} = 6\,\text{m}$	1
27		$v = f\lambda$ (1) $\lambda = \dfrac{3 \times 10^8}{2100 \times 10^6}$ (1) $= 0 \cdot 14\,\text{m}$ (1)	3
28		 Same wavelength after gap (1) Circular waves after gap (1)	2

Question		Answer		Max. mark
29		P microwaves Q infrared R ultraviolet (2 marks for all three correct; 1 mark for one or two correct.)		2
30	a)	*For showing refraction towards normal in glass* (1) *For showing refraction away from normal when emerging from glass into air* (1)		2
	b)	*For correct identification of an angle of refraction* (1)		1
31		Source Z is most suitable (1) Only gamma radiation would penetrate the steel. A long half-life would mean that the source could be used repeatedly for years without replacement. (1)		2
32		$A = \dfrac{N}{t}$ (1) $= \dfrac{4 \cdot 14 \times 10^8}{5 \times 60 \times 60}$ (1) $= 2 \cdot 3 \times 10^4 \, \text{Bq}$ (1)		3
33	a)	$D = \dfrac{E}{m}$ (1) $= \dfrac{8 \cdot 2 \times 10^{-3}}{90}$ (1) $= 9 \cdot 1 \times 10^{-5} \, \text{Gy}$ (1)		3
	b)	$H = D w_r$ (1) $= 9 \cdot 1 \times 10^{-5} \times 1$ (1) $= 9 \cdot 1 \times 10^{-5} \, \text{Sv}$ (1)		3
34		2 days $= (4 \times 12)$ hours $= 4$ half-lives (1) $128 \rightarrow 64 \rightarrow 32 \rightarrow 16 \rightarrow 8$, so activity $= 8 \, \text{MBq}$ (1)		2

Extended questions

Question			Answer		Max. mark
1	**a)**		$W = mg$	**(1)**	**3**
			$= 9 \cdot 7 \times 10^7 \times 9 \cdot 8$	**(1)**	
			$= 9 \cdot 5 \times 10^8 \text{ N}$	**(1)**	
	b)	**(i)**			**2**
			$R^2 = (1 \cdot 8 \times 10^3)^2 + (4 \cdot 6 \times 10^3)^2$	**(1)**	
			Resultant force $= 4 \cdot 9 \times 10^3 \text{ N}$	**(1)**	
		(ii)	$\tan X = \dfrac{4 \cdot 6 \times 10^3}{1 \cdot 8 \times 10^3}$	**(1)**	**2**
			Bearing of 111° (or 21° South of East)	**(1)**	
					(7)
2	**a)**	**(i)**	$a = \dfrac{v - u}{t}$	**(1)**	**3**
			$= \dfrac{0 - 16}{30}$	**(1)**	
			$= -0 \cdot 53 \text{ ms}^{-2}$	**(1)**	
		(ii)	Cyclist is moving at constant speed.		**1**
		(iii)			**2**
	b)	**(i)**	Total distance = area under speed–time graph.	**(1)**	**3**
			$= \dfrac{1}{2} \times 25 \times 16 + 195 \times 16 + \dfrac{1}{2} \times 30 \times 16$	**(1)**	
			$= 3560 \text{ m}$	**(1)**	
		(ii)	$\bar{v} = \dfrac{d}{t}$	**(1)**	**3**
			$= \dfrac{3560}{250}$	**(1)**	
			$= 14 \cdot 24 \text{ ms}^{-1} = 14 \cdot 2 \text{ ms}^{-1}$	**(1)**	
					(12)

Question			Answer		Max. mark
3	a)		$a = \dfrac{v-u}{t}$ (1) $9 \cdot 8 = \dfrac{v-0}{0 \cdot 55}$ (1) $v = 5 \cdot 39 \, \text{ms}^{-1}$ (1)		3
	b)		$\bar{v} = \dfrac{u+v}{2} = \dfrac{0+5 \cdot 39}{2} = 2 \cdot 7 \, \text{ms}^{-1}$ (1) $\bar{v} = \dfrac{d}{t}$ (1) $2 \cdot 7 = \dfrac{d}{0 \cdot 55}$ (1) $d = 1 \cdot 49 \, \text{m}$ (1)		4
	c)		Time for ball to fall to ground is the same as in part **a)**.		1
					(8)
4	a)	(i)	Mercury and Mars		1
		(ii)	Saturn and Jupiter		1
		(iii)	Neptune		1
	b)	(i)	Ion drive engines expel a beam of gas ions which produces a small unbalanced force on the spacecraft. (1) When applied over a long period of time, the spacecraft would reach a very high velocity to travel the vast distances. (1)		2
		(ii)	Gravity assist allows a spacecraft to increase its speed by 'flying by' a larger object (e.g. a planet) in space which is orbiting the Sun. When the spacecraft approaches the planet, it will start to orbit the planet. (1) When the spacecraft leaves the orbit, some of the kinetic energy of the planet is transferred to the spacecraft and its velocity increases. (1)		2
	c)		Solar cells are used to transform solar energy into electrical energy and usually recharge batteries for use when the cells are shaded from the Sun. Using current technology, solar cells are practical for spacecraft orbiting up to the orbit of Mars. One suggestion is to increase the size of the solar cell array as the spacecraft travels further away from the Sun.		1
					(8)
5	a)		$I = \dfrac{1 \cdot 5}{3}$ $= 0 \cdot 5 \, \text{A}$ (1)		1
	b)		$P = IV$ (1) $= 0 \cdot 5 \times 4 \cdot 5$ (1) $= 2 \cdot 25 \, \text{W}$ (1)		3
	c)		The ammeter reading will increase. (1) Adding lamp in parallel reduces total resistance so current increases. (1)		2
					(6)

Question			Answer		Max. mark
6	a)	(i)	$V_R = V_s - V_{LDR} = 3 \cdot 0\,V$	(1)	4
			$\dfrac{V_{LDR}}{V_R} = \dfrac{R_{LDR}}{R}$	(1)	
			$\dfrac{2}{3} = \dfrac{R_{LDR}}{2250}$	(1)	
			$R_{LDR} = 1500\,\Omega$	(1)	
		(ii)	140 units		1
	b)	(i)	As R_{LDR} increases, V_{LDR} increases.	(1)	3
			When V_{LDR} reaches switching voltage, the MOSFET turns on.	(1)	
			Current is now in the relay which completes the lamp circuit.	(1)	
		(ii)	Lamps will switch on at a lower outside light level.	(1)	3
			Resistance of LDR must be greater than $1500\,\Omega$.	(1)	
			to maintain the MOSFET switching voltage.	(1)	
					(11)
7	a)		$E_h = cm\Delta T$	(1)	3
			$\quad = 1800 \times 2 \cdot 8 \times 150$	(1)	
			$E = 756\,000\,J$	(1)	
	b)		$P = \dfrac{E_h}{t}$	(1)	3
			$1500 = \dfrac{756\,000}{t}$	(1)	
			$t = 504\,s$	(1)	
	c)	(i)	Heat energy loss from the appliance to the surroundings means it will take longer to supply the energy required to heat the oil to 170°C.		1
		(ii)	Use a cover/lid.		1
					(8)
8	a)		$T_1 = 20\,°C = 293\,K$		3
			$T_2 = 28\,°C = 301\,K$		
			$\dfrac{P_1}{T_1} = \dfrac{P_2}{T_2}$	(1)	
			$\dfrac{100}{293} = \dfrac{P_2}{301}$	(1)	
			$p_2 = 103\,kPa$	(1)	
	b)		As temperature increases, the average E_k of the gas particles increases	(1)	3
			so the particles collide with the walls of the container more frequently with greater force	(1)	
			so pressure increases.	(1)	
	c)		Place temperature sensor inside the flask to measure gas temperature directly.		1
					(7)

Question			Answer		Max. mark
9	a)		Refraction occurs when light travels from one medium into another with a change in the wave speed and wavelength.		1
	b)	(i) and (ii)	For showing refraction towards normal in glass	(1)	3
			For showing refraction away from normal when emerging from glass into air	(1)	
			For correct identification of one angle of incidence	(1)	
10	a)	(i)	X: infrared	(1)	2
			Y: ultraviolet	(1)	
		(ii)	$v = \dfrac{d}{t}$	(1)	3
			$3 \times 10^8 = \dfrac{6 \cdot 7 \times 10^{11}}{t}$	(1)	
			$t = 2 \cdot 2 \times 10^3 \, \text{s}$	(1)	
					(9)
11	a)		$T = \dfrac{1}{f}$	(1)	3
			$4 \cdot 0 \times 10^{10} = \dfrac{1}{f}$	(1)	
			$f = 2 \cdot 5 \times 10^9 \, \text{Hz}$	(1)	
	b)		$v = f\lambda$	(1)	3
			$3 \times 10^8 = 2 \cdot 5 \times 10^9 \times \lambda$	(1)	
			$\lambda = 0 \cdot 12 \, \text{m}$	(1)	
					(6)
12	a)	(i)	Half-life is the time for the activity of a radioactive substance to reduce to half of its original value. OR The time taken for one half of the nuclei of a radioactive substance to disintegrate.		1
		(ii)	12 hours		1
		(iii)	$80 \rightarrow 40 \rightarrow 20 \rightarrow 10$	(1)	2
			$10 \, \text{kBq}$	(1)	

Question			Answer		Max. mark
	b)	(i)	$H = Dw_r$	(1)	3
			$= 15 \times 2$	(1)	
			$= 30\,\mu Sv$	(1)	
		(ii)	$\dot{H} = \dfrac{H}{t}$	(1)	3
			$= \dfrac{30}{3}$	(1)	
			$= 10\,\mu Svh^{-1}$	(1)	
					(10)

Experimental and data-handling questions

Question			Answer		Max. mark
1	a)		$v = \dfrac{d}{t}$	(1)	3
			$= \dfrac{50 \times 10^{-3}}{0 \cdot 085}$	(1)	
			$= 0 \cdot 59\ ms^{-1}$	(1)	
	b)		Any value greater than zero but less than answer to part **a)**. A value must be given and unit is required.		1
2			Heating tank B has the best insulation.	(1)	2
			Tank B takes longer than tank A for the temperature to fall to 40 °C.	(1)	
3	a)	(i)	Resultant force causing acceleration $F_R = ma$	(1)	3
			$= 0 \cdot 4 \times 45$	(1)	
			$= 18\,N$	(1)	
		(ii)	Acceleration increases.	(1)	2
			As fuel is used, the mass of the rocket reduces.		
			The acceleration increases because the engine force is constant.	(1)	
	b)		Any value (in minutes) greater than 99 minutes but less than 847 minutes. A value must be given and unit is required.		1
4	a)				2
			For correct symbols	(1)	
			For correct placement of meters	(1)	
	b)		The power developed in the resistor $P = IV$	(1)	3
			$P = 0 \cdot 60 \times 12 = 7 \cdot 2\,W$	(1)	
			This is greater than the 5 W labelled power rating so the resistor overheats.	(1)	

Question			Answer		Max. mark
5	a)	(i)	$I = 0 \cdot 075\,\text{A}$	(1)	4
			$V = IR$	(1)	
			$4 \cdot 2 = 0 \cdot 075 \times R$	(1)	
			$R = 56\,\Omega$	(1)	
		(ii)	Resistance stays the same	(1)	2
			because the graph is a straight line through the origin.	(1)	
	b)		$\dfrac{1}{R_{\text{T}}} = \dfrac{1}{R_1} + \dfrac{1}{R_2}$		3
			$= \dfrac{1}{33} + \dfrac{1}{56}$	(1)	
			$= 0 \cdot 048$	(1)	
			$R_{\text{T}} = 20 \cdot 76 = 21\,\Omega$	(1)	
6	a)		$P \times V = 2000, 1995, 2002, 2001$	(1)	2
			$P \times V = \text{constant}$	(1)	
	b)		Gas molecules collide with walls of container more often	(1)	3
			so average force increases	(1)	
			causing an increase in pressure.	(1)	
	c)		To reduce the inaccuracy of the syringe volume, since the volume of air contained in the tubing is not measured.		1
7	a)		<table><tr><td>*Transverse*</td><td>*Longitudinal*</td></tr><tr><td>light wave radio wave</td><td>sound wave</td></tr></table>		2
	b)	(i)	Transverse because particles vibrate at right angles to the direction of energy transfer of the wave.		1
		(ii)	$f = \dfrac{N}{t} = \dfrac{10}{5} = 2\,\text{Hz}$	(1)	4
			$v = f\lambda$	(1)	
			$= 2 \times 2$	(1)	
			$v = 4\,\text{ms}^{-1}$	(1)	
8	a)		The source could be emitting gamma radiation, which would not be absorbed by the aluminium sheet.		1
	b)		The technician could compare the reduced count rate due to the aluminium sheet with the background count rate.		1
			If the reduced count rate is greater than background count rate then the source is emitting gamma radiation.		
	c)		Handle the sources with tongs.		1
			Never point the sources at the eyes.		

Open-ended questions

Question	Answer	Max. mark
1	▶ Cars have to speed up regularly; reducing the overall mass of the car would reduce the unbalanced force required for acceleration ($F = ma$), reducing fuel needed. The mass of the car will affect the fuel consumption. ▶ Making the car more streamlined would reduce the frictional forces acting on the car and so less fuel would be needed to provide the reduced forward engine force. ▶ Manufacturers produce cars that transform their kinetic energy into electrical energy when braking, to be stored in rechargeable cells. These cells can provide an additional energy source to energise a motor, which assists the car's movement. ▶ Modern totally electric cars use only electrical energy from a battery, which is replaced by recharging the battery using electricity from the National Grid. This electricity is generated more efficiently than the efficiency of conventional fuel cars and represents a more efficient method of driving. ▶ On the occasions when cars are in long queues or traffic jams, energy is used up while the car engine idles. This is common on motorways and in towns. Some manufacturers have already introduced a detector to switch the engine off and to re-start when the driver is able to drive ahead.	3
2	▶ If air resistance is ignored, then at the planet surface all objects fall with the **same** acceleration due to gravity, g, so objects will take the same time to fall equal distances. ▶ Different planets have different values for this acceleration, g. The time taken to fall the same distance would be longer on planets where g is smaller. ▶ On Earth, air resistance could reduce the acceleration of a falling object. If objects were falling far enough through air to reach terminal velocity, then this velocity would be different for objects with less streamlining, and affect the time taken to reach the ground. ▶ The acceleration also depends on the height above the planet. Objects released at higher altitudes would take longer to fall the same distance than objects released closer to the Earth's surface because g is less.	3
3	▶ As the rocket rises, fuel is used up and the rocket mass reduces. This reduces the rocket weight and so the unbalanced upward force increases, causing acceleration to increase, as $a = \dfrac{F}{m}$. ▶ As the rocket rises, the gravitational field strength decreases. This causes the rocket weight to reduce and so the unbalanced upward force increases, causing acceleration to increase, as $a = \dfrac{F}{m}$. ▶ As the rocket rises, the frictional force reduces because of the atmosphere being less dense at altitude. This reduces air resistance and so the unbalanced upward force increases, causing acceleration to increase, as $a = \dfrac{F}{m}$.	3
4	▶ Assume that the average mass of a student is 60 kg. ▶ Student has a weight of $W = mg = 588\,\text{N}$. ▶ The total weight of the student will be exerted on the ground where their feet are in contact. ▶ Assume that the area of one foot in contact with ground is $0{\cdot}2\,\text{m} \times 0{\cdot}08\,\text{m} = 0{\cdot}016\,\text{m}^2$. ▶ Total area in contact $= 0{\cdot}032\,\text{m}^2$. ▶ $\text{Pressure} = \dfrac{\text{force}}{\text{area}} = \dfrac{588}{0{\cdot}032} = 18375\,\text{Pa} = 18000\,\text{Pa}$.	3

Question		Answer	Max. mark
5		▶ The final temperature will be between both starting temperatures. ▶ The heat energy E_h lost by the copper will be gained by the water. ▶ From $\Delta T = \dfrac{E_h}{cm}$ the rise in temperature of the water and fall in temperature of the copper will depend on the respective masses of water and copper. ▶ Specific heat capacity for water is much greater than for copper, so if masses were equal, final temperature would be closer to water's starting temperature.	3
6		▶ Astronomical objects in space, like stars and galaxies, emit electromagnetic radiation across all bands of the electromagnetic spectrum. Different detectors are required to receive signals from these different bands. Radiation from some bands of the electromagnetic spectrum is absorbed by the Earth's atmosphere, so some telescopes need to be located in satellites above the atmosphere to detect this radiation. ▶ Telescopes contain different detectors to receive the signals of each particular band in the electromagnetic spectrum. For example: 　◆ Radio telescopes detect radio waves, which can be used to help produce maps of the positions of astronomical objects is space. 　◆ Infrared telescopes can be used to detect infrared radiation from stars and galaxies which are obscured behind dense regions of dust or gas and which do not allow visible light to pass through. This has led to the discovery of a star-forming region at the centre of our galaxy, the Milky Way. 　◆ Visible light from stars and galaxies can be analysed to obtain line spectra which can identify the elements present. 　◆ Gamma rays detected in space have resulted in data from stars exploding or colliding stars and black holes to be discovered.	3

Scientific literacy questions

Question		Answer	Max. mark
1	a)	Neutron stars are thought to be formed when large stars collapse.	1
	b)	During the collapsing process, electrons and protons combine to form neutrons.	1
	c)	$T = \dfrac{1}{f}$ (1) $T = \dfrac{1}{716}$ (1) $T = 1 \cdot 4 \times 10^{-3}\ \text{Hz}$ (1)	3
2	a)	It is thought that the asteroid belt was formed about 4·5 billion years ago.	1
	b)	The gravitational effects from the larger planets – Jupiter, Saturn, Neptune or Uranus – can cause some asteroids to alter their orbit.	1
	c)	$d = vt$ (1) $d = 3 \times 10^8 \times 4 \cdot 2 \times 10^{-5} \times 365 \cdot 25 \times 24 \times 60 \times 60$ (1) $\quad = 3 \cdot 98 \times 10^{11}\,\text{m} = 4 \cdot 0 \times 10^{11}\,\text{m}$ (1)	3

Section 1

> **Total marks:** 25
>
> Attempt ALL questions.
>
> The answer to each question is **either** A, B, C, D or E. Decide what your answer is, then circle the appropriate letter. There is **only one correct** answer to each question.
>
> Reference may be made to the Data sheet and to the Relationships sheet.
>
> Allow yourself around 30 minutes for Section 1.

	MARKS	STUDENT MARGIN

1 Which of the following contains two scalar quantities and one vector quantity?   **1**   Dynamics 1

   A displacement, velocity, acceleration

   B speed, velocity, displacement

   C time, distance, force

   D acceleration, mass, displacement

   E displacement, force, velocity

2 An athlete sprints 50 m South then 30 m North in 8 seconds.   **1**   Dynamics 1

Which row in the table shows the average speed and average velocity of the athlete?

	Average speed ($m\,s^{-1}$)	Average velocity ($m\,s^{-1}$)
A	2·5	2·5 North
B	2·5	10 South
C	10	2·5 North
D	10	2·5 South
E	10	10 South

3 A ball is placed on a slope.   **1**   Dynamics 3

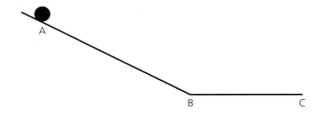

The ball is released from rest at point A and rolls down the slope. The ball takes 5 s to reach point B and has a speed of $4\,m\,s^{-1}$.

The acceleration of the ball between A and B, in ms^{-2}, is

   A 0·8

   B 1·25

   C 9·0

   D 10·0

   E 20·0.

	MARKS	STUDENT MARGIN

4 A 1 kg ball is dropped into a deep well. The graph shows the speed of the ball from the instant of its release in air until it has fallen several metres through the water towards the bottom of the well.

1   Dynamics 4

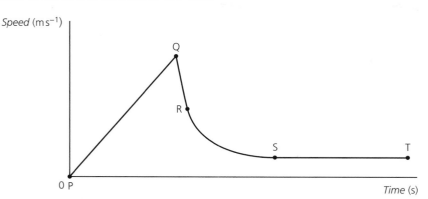

Which part of the graph shows when the ball is experiencing a constant downward force of 9·8 N?

A   PQ only

B   QR only

C   RS only

D   ST only

E   PT.

5 Two identical balls, X and Y, are projected horizontally from the edge of a cliff.
The path taken by each ball is shown.
The effects of air resistance can be ignored.

1   Dynamics 6

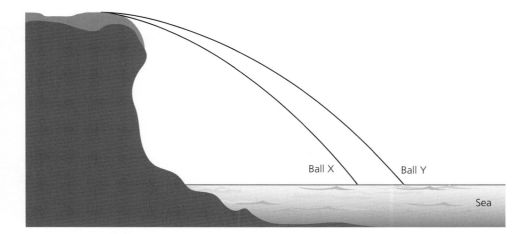

A student makes the following statements about the motion of the two balls.

I   They take different times to reach sea level.

II   They have a horizontal acceleration.

III   They have different horizontal velocities.

Which of these statements is/are correct?

A   I only

B   II only

C   III only

D   I and II only

E   I and III only.

6 Some astronomical objects that exist in the universe are described in the table. Which row most accurately describes these objects?

	Planet	Star	Exoplanet
A	a star and the objects that orbit it	a solid object in space	all of the energy and matter in existence
B	a collection of stars	a natural satellite that orbits a planet	a mass of hot gas that emits heat and light
C	a universe	all of the energy and matter in existence	a collection of stars
D	a natural satellite that orbits a star	a mass of hot gas that emits heat and light	a planet outside our solar system orbiting another star
E	a mass of hot gas that emits heat and light	a planet outside our solar system orbiting another star	a star and the objects that orbit it

1   Space 1

7 The Jason-3 satellite, which monitors the surface of the oceans, has a period of 112 minutes and an orbital altitude of 1340 km.

The television communications satellite Intelsat-18 has a period of 1440 minutes and an orbital altitude of 35 900 km.

Which of the following gives the period of a satellite that has an orbital altitude of 20 000 km?

- **A**   82 minutes
- **B**   94 minutes
- **C**   720 minutes
- **D**   1450 minutes
- **E**   1740 minutes.

1   Space 1

8 The Big Bang theory about the origin of the universe suggests that the Big Bang happened approximately

- **A**   14 thousand years ago
- **B**   1·4 million years ago
- **C**   14 million years ago
- **D**   1·4 billion years ago
- **E**   14 billion years ago.

1   Space 2

9 The unit of current is the ampere.

One ampere can also be expressed as

- **A**   one ohm per volt
- **B**   one joule per coulomb
- **C**   one joule per second
- **D**   one coulomb per second
- **E**   one volt per joule.

1   Electricity 1

10 The circuit shows a 60 V supply connected across two resistors.

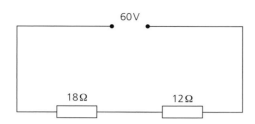

The total charge passing through the 12 Ω resistor in 2 minutes is

A   4 C
B   24 C
C   60 C
D   240 C
E   1440 C.

11 The diagrams show the electric field lines between two oppositely charged points. Which diagram shows the correct direction of travel of a proton and an electron placed in the positions shown in the electric field?

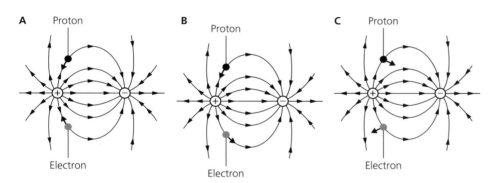

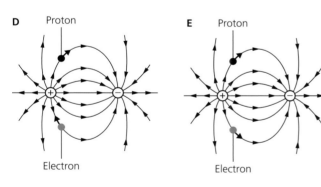

	MARKS	STUDENT MARGIN

12 The graph shows how the current is related to the applied potential difference for two separate resistors P and Q.

1 — Electricity 3

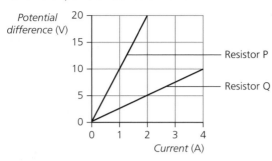

Which row in the table shows the values of resistor P and resistor Q?

	Resistance of P (Ω)	Resistance of Q (Ω)
A	0·1	0·4
B	10	2·5
C	5	5
D	20	10
E	10	20

13 Which is the correct symbol for a light-dependent resistor (LDR)?

1 — Electricity 4

A

B

C

D

E
t

14 A circuit is set up as shown.

1 — Electricity 4

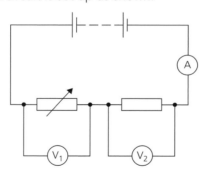

The resistance of the variable resistor is now decreased.

Which row in the table correctly shows the effect on the readings on the ammeter and voltmeters?

	Reading on voltmeter V_1	Reading on voltmeter V_2	Reading on ammeter
A	decreases	decreases	decreases
B	increases	unchanged	increases
C	increases	unchanged	decreases
D	decreases	decreases	increases
E	decreases	increases	increases

	MARKS	STUDENT MARGIN

15 A solid substance of mass 0·25 kg is placed in an insulated container and heated continuously by a 200 W heater.

The graph shows how the temperature of the substance changes over time.

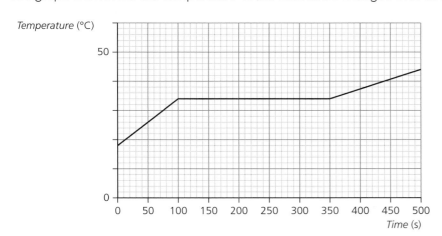

The specific latent heat of fusion of the substance is

A   5·0 J kg^{-1}

B   8·0 × 10^4 J kg^{-1}

C   1·2 × 10^5 J kg^{-1}

D   2·0 × 10^5 J kg^{-1}

E   4·0 × 10^4 J kg^{-1}.

16 Which graph represents the correct relationship between the pressure p and the volume V of a fixed mass of gas at constant temperature?

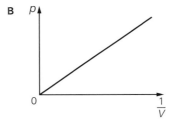

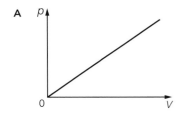

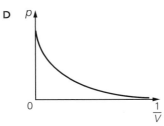

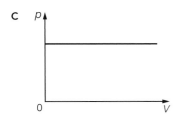

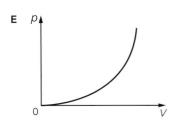

15 — 1 — Properties of matter 2

16 — 1 — Properties of matter 3

| | MARKS | STUDENT MARGIN |

17 The diagram gives information about a wave.

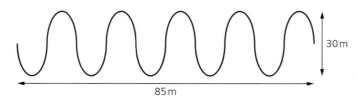

30 m

85 m

MARKS 1   Waves 1

The time taken for the wave to travel 85 m is 2·5 s.

The following statements are made about the wave.

I   The speed of the wave is 34 m s⁻¹.

II   The frequency of the wave is 2 Hz.

III   The amplitude of the wave is 30 m.

Which of these statements is/are correct?

A   I only

B   II only

C   I and II only

D   I and III only

E   I, II and III.

18 A ray of white light passes from air into glass.

MARKS 1   Waves 3

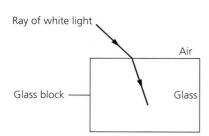

Ray of white light

Air

Glass block

Glass

Which row in the table correctly shows the effect on the wave speed, frequency and wavelength of the light when it enters the glass?

	Wave speed	Frequency	Wavelength
A	decreases	increases	constant
B	increases	constant	decreases
C	constant	increases	increases
D	decreases	constant	decreases
E	decreases	increases	constant

| | MARKS | STUDENT MARGIN |

19 Which row in the table describes alpha (α), beta (β) and gamma (γ) radiations?

	α	β	γ
A	helium nucleus	electron from the nucleus	electromagnetic radiation
B	electron from the nucleus	helium nucleus	electromagnetic radiation
C	electromagnetic radiation	electron from the nucleus	helium nucleus
D	electromagnetic radiation	helium nucleus	electron from the nucleus
E	helium nucleus	electromagnetic radiation	electron from the nucleus

20 A radioactive source emits alpha, beta and gamma radiation. A detector, connected to a counter, is placed 10 mm in front of the source. The counter records 500 counts per minute.

A 5 mm sheet of aluminium is placed between the source and the detector. The counter now records 50 counts per minute.

The radiation now detected by the detector is

A alpha only

B alpha and beta only

C beta only

D beta and gamma only

E gamma only.

21 For a particular radioactive source, 30 000 atoms decay in a time of 5 minutes.

The activity of this source is

A $1\cdot7 \times 10^{-4}$ Bq

B $0\cdot01$ Bq

C 100 Bq

D $1\cdot5 \times 10^{5}$ Bq

E $9\cdot0 \times 10^{6}$ Bq.

22 A tissue sample of mass 0·08 kg is exposed to 2 mJ of beta radiation.

The absorbed dose received by the tissue is

A $0\cdot025$ Gy

B $25\cdot0$ Gy

C $40\cdot0$ Gy

D $160\cdot0$ Gy

E $2\cdot5 \times 10^{4}$ Gy.

23 A sample of tissue receives an absorbed dose of 50 mGy when exposed to slow neutrons.

The equivalent dose is

A $1\cdot5 \times 10^{-5}$ Sv

B $0\cdot15$ Sv

C $0\cdot5$ Sv

D 150 Sv

E 500 Sv.

24 When considering the exposure of humans to radiation, the *effective equivalent dose* is used to assess the potential for long-term effects of radiation exposure that might occur in the future.

Which row in the table correctly shows the current annual safe limits of effective equivalent dose for members of the public and for employees?

	Members of public and employees under 18	Employees over 18
A	20 mSv	1 mSv
B	100 mSv	20 mSv
C	20 mSv	100 mSv
D	1 mSv	1 mSv
E	1 mSv	20 mSv

1 Radiation 1

25 Radioactive tracers are liquids that are injected into patients to check the health and function of various organs in the body.

The tracer should

A emit gamma radiation and have a long half-life

B emit alpha radiation and have a short half-life

C emit gamma radiation and have a short half-life

D emit alpha radiation and have a long half-life

E emit beta radiation and have a long half-life.

1 Radiation 1

Section 2

Total marks: 110

Attempt ALL questions.

Write your answers clearly in the spaces provided. If you need additional space for answers or rough work, please use separate pieces of paper.

Reference may be made to the Data sheet and to the Relationships sheet.

Allow yourself around 2 hours for Section 2.

MARKS   STUDENT MARGIN

1   A passenger aircraft of mass 360 000 kg prepares for take-off.

The speed–time graph for the aircraft's motion on the runway from rest until it takes off is shown.

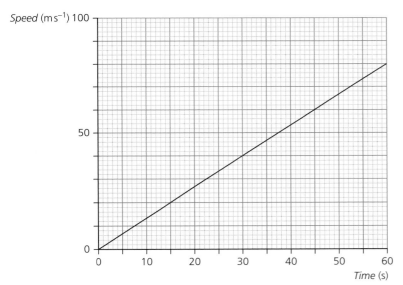

a)   (i)   Calculate the acceleration of the aircraft during take-off.

Space for working and answer

3   Dynamics 3

(ii) The forward force produced by the aircraft engines is 500 kN.
Calculate the average frictional force acting on the aircraft during take-off.

Space for working and answer

4   Dynamics 4

(iii) Calculate the length of runway required by the aircraft for take-off.

Space for working and answer

3   Dynamics 2

b) During the flight, the aircraft flies at a constant speed and height.

Calculate the upward force acting on the aircraft.

Space for working and answer

3   Dynamics 4

c) When flying an aircraft between London and New York, an airline pilot is exposed to cosmic radiation at an equivalent dose rate of 8 μSv h^{-1}. Each flight lasts 7 hours. The pilot makes 106 of these flights in one year.

Calculate the equivalent dose received by the pilot from this exposure in one year.

Space for working and answer

3   Radiation 1

	MARKS	STUDENT MARGIN

2 a) A city in Europe is famous for its water fountain situated in a lake.

Every second, water pumps force 500 kg of water out of a vertical nozzle at a speed of 55 m s^{-1}.

Calculate the kinetic energy of this water as it emerges from the nozzle. **3** Dynamics 5

Space for working and answer

b) Show that the maximum height that the 500 kg of water could reach is 155 m. **3** Dynamics 5

Space for working and answer

c) In practice, the water jet reaches a height of 140 m instead of 155 m.

State the reason for this difference. **1** Dynamics 5

3 Sir Isaac Newton described three laws of motion.

For each of these laws, describe an example of the law in use during everyday life.

3	Dynamics 4

4 Read the passage below and answer the questions that follow.

Sunspots

Sunspots are the dark spots that appear on the surface of the Sun. They are caused by intense magnetic fields appearing beneath the Sun's surface. These magnetic fields disrupt the natural convection within the Sun, causing a reduction in the surface temperature at that point, which appears as a dark spot.

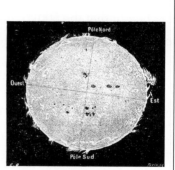

Sunspots release huge amounts of energy known as 'solar flares'. These flares lead to the ejection of billions of tonnes of high-energy solar particles. Electromagnetic (EM) radiation is also emitted. Such releases are known as coronal mass ejections (CME). The particles and energy travel into space.

High-energy particles emitted by solar flares rarely reach Earth and, when they do, its magnetic field deflects most of them. Background radiation levels are not significantly changed by the particles which penetrate the atmosphere.

However, satellites in the path of these particles can suffer damage to sensitive equipment on board. Early warning of the arrival of solar flares allows operators to place satellites into 'protective mode'.

The arrival of high electromagnetic energy, in particular ultraviolet radiation and X-rays, can also cause significant damage. When this EM radiation reaches the Earth, it can cause changes to the upper atmosphere, which can affect the orbits of low-altitude satellites. The EM radiation can also cause localised disturbances of the Earth's magnetic field. On one occasion, such interference was claimed to be responsible for causing power failures in Canada which affected millions of households for several hours.

The NASA space agency monitored sunspots, using their two satellite-based Solar Terrestrial Relations Observatories (STEREO A and B). One particular observation from these satellites recorded a solar flare ejected from the Sun which took 2·3 days to reach Mars, a distance of $2·28 \times 10^{11}$ metres from the Sun.

Sunspot activity increases to a maximum in 11-year cycles. There is ongoing research to identify a connection between solar activity and our terrestrial climate. Currently, scientists cannot predict when a sunspot or solar flare will appear, but detection and early warning techniques have improved.

	MARKS	STUDENT MARGIN

a) What causes sunspots to appear?

MARKS: 1 — STUDENT MARGIN: Space 1

b) Which parts of the electromagnetic spectrum can cause damage to satellites?

MARKS: 2 — STUDENT MARGIN: Waves 2

c) Calculate the average speed, in ms^{-1}, of the solar flare which reached Mars.

Space for working and answer

MARKS: 3 — STUDENT MARGIN: Dynamics 1

5 The exploration vehicle Curiosity landed on the planet Mars in 2012.

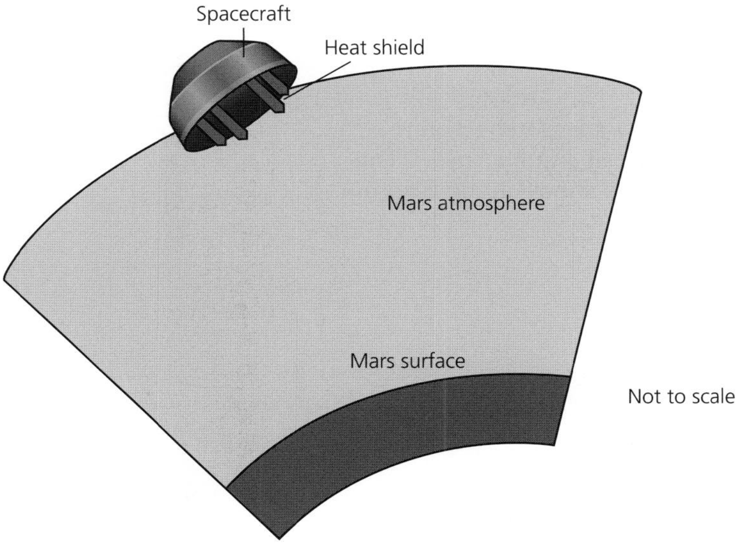

Spacecraft
Heat shield
Mars atmosphere
Mars surface
Not to scale

A spacecraft containing Curiosity entered the planet's atmosphere 125 km from its surface at a speed of 5900 ms^{-1}, and decelerated to a speed of around 400 ms^{-1} at 11 km from the surface.

The average frictional force acting on the spacecraft was 3.7×10^5 N.

a) (i) Calculate the work done by the frictional force on the spacecraft during this time.

Space for working and answer

MARKS: 3 — STUDENT MARGIN: Dynamics 5

MARKS   STUDENT MARGIN

(ii) Explain why a heat shield on the spacecraft was necessary for this part of the descent.

1   Space 1

(iii) At 11 km from the surface, the spacecraft deployed a parachute to increase its deceleration.

State why parachutes would have been ineffective to decelerate any spacecraft landing on the Moon.

1   Space 1

b) Closer to the surface of Mars, the spacecraft containing Curiosity shed its parachute, heatshield and outer shell.

The weight of the spacecraft at this stage was now 12210 N.

(i)   Show that the mass of the spacecraft at this stage was 3300 kg.

Space for working and answer

2   Space 1

(ii) At 1·6 km from the surface, thruster engines were switched on to decelerate the spacecraft to a landing speed of around 1 m s^{-1}. The forces acting on the spacecraft at this time are shown in the diagram.

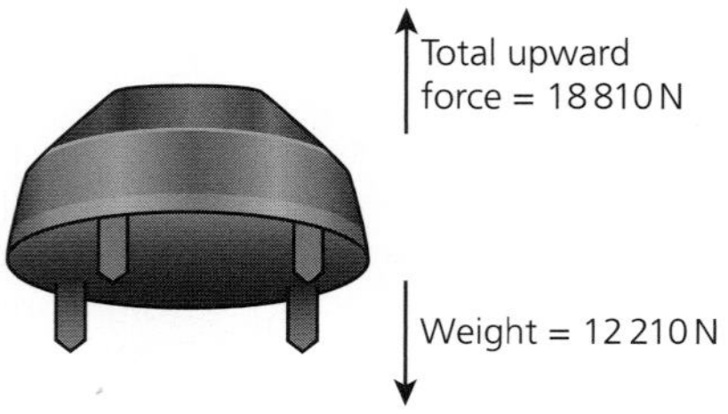

Total upward force = 18 810 N

Weight = 12 210 N

Calculate the magnitude of the deceleration of the spacecraft.

Space for working and answer

4   Space 1

MARKS   STUDENT MARGIN

6 A student decides to construct an electronic egg timer.
The student connects a capacitor in the following circuit.

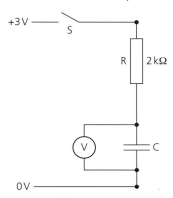

The voltage across the capacitor is measured by a voltmeter.

When switch S is open the voltmeter reading is zero. The switch is now closed.
A graph of the voltmeter readings against time is shown.

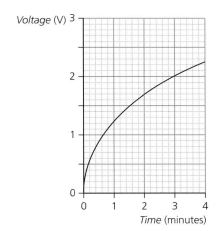

a) **(i)** State the voltage across the capacitor after 3 minutes.   1   Electricity 4

(ii) Explain why the current in R is not constant after switch S is closed.   1   Electricity 4

(iii) Calculate the current in R after 3 minutes.   4   Electricity 3
Space for working and answer

MARKS   STUDENT MARGIN

b) The circuit is now connected to a switching circuit to operate a buzzer.

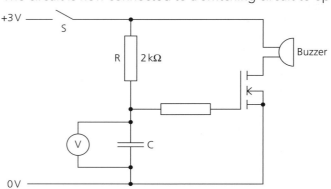

Explain how the circuit operates to make the buzzer sound.

2   Electricity 4

7 The following circuit is constructed to operate two identical 12 V, 18 W lamps from a 30 V supply.

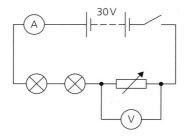

a) When the switch is closed, the lamps operate at their correct power rating. Calculate:

(i) the reading on the ammeter;

Space for working and answer

3   Electricity 5

(ii) the reading on the voltmeter.

Space for working and answer

1   Electricity 4

	MARKS	STUDENT MARGIN

b) A new circuit is constructed with a 12V supply and the same lamps. The resistance of each lamp is 8 Ω. The variable resistor is set to a resistance of 16Ω.

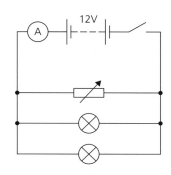

(i) Calculate the combined resistance of the circuit.

Space for working and answer

        3   Electricity 4

(ii) The resistance of the variable resistor is now increased. Explain what happens to the reading on the ammeter.

        2   Electricity 4

8 A house has a solar heat exchanger and a solar cell array installed on its roof.

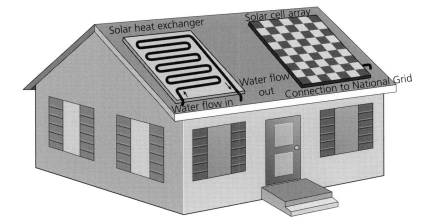

Solar heat exchanger	Solar cell array
Water is pumped into the solar heat exchanger, where heat energy from the Sun raises its temperature. The heated water continues to a separate heat exchanger inside the house where the energy is stored for later use.	The solar cell array converts solar energy into electrical energy where it is connected and supplied to the National Grid.
	The householder is paid for the electrical energy supplied.

MARKS | STUDENT MARGIN

The householder decided to compare the outputs of the solar heat exchanger and the solar cell array.

The solar heat exchanger and solar cell array were tested over 5 hours on a sunny day. The following results were recorded.

Solar heat exchanger	
Average water temperature **in** (°C)	15·5
Average water temperature **out** (°C)	25·8
Mass of water heated during 5 hours (kg)	250·0

Solar cell array	
Average output power during 5 hours (W)	350

a) Calculate the solar energy transferred into heat energy by the solar heat exchanger during the 5 hours.

Space for working and answer

3 | Properties of matter 1

b) Calculate how long it would take, in hours, for the solar cell array to supply the equivalent amount of heat energy produced by the solar heat exchanger in part **a)**.

Space for working and answer

3 | Electricity 5

9 A student sets up an experiment to calculate a value for the specific latent heat of vaporisation of water. The apparatus used in the experiment is shown.

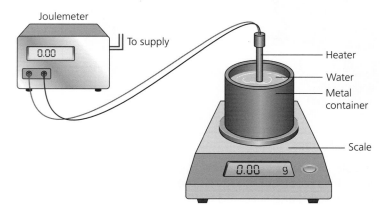

The heater is switched on and the water in the metal container is heated.
The student eventually sees the reading of the mass on the scales begin to reduce.

| | | MARKS | STUDENT MARGIN |

a) Why does the mass reading on the scales start to reduce?

MARKS: 1 — Properties of matter 2

b) When the mass reading on the scales start to reduce, the student records initial and final readings from the scales and the joulemeter.

Use the student's results to determine a value for the specific latent heat of vaporisation of water.

MARKS: 4 — Properties of matter 2

	Scales (g)	Joulemeter ($J \times 10^5$)
Initial reading	1710	7·52
Final reading	1460	1·62

Space for working and answer

c) The student notices that the value calculated for the specific latent heat of vaporisation for water is different from the accepted value.

Suggest a modification which would improve the accuracy of the experiment.

MARKS: 1 — Dynamics 5

10 A car engine manufacturer tests a diesel engine. Diesel engines have cylinders filled with air. A piston is used to compress the air, causing an increase in its temperature.

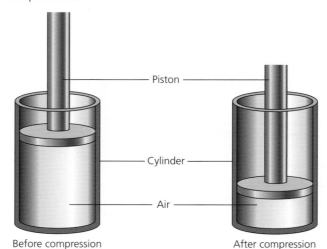

Before compression   After compression

During one test the manufacturer obtained the following results.

	Before compression	After compression
Volume of air in cylinder (m³)	$5 \cdot 0 \times 10^{-4}$	$2 \cdot 0 \times 10^{-5}$
Pressure of air in cylinder (Pa)	$1 \cdot 0 \times 10^{5}$	$5 \cdot 0 \times 10^{6}$
Temperature of air in cylinder (°C)	$20 \cdot 0$	

a) Calculate the final temperature of the air in the cylinder after compression.

Space for working and answer

3   Properties of matter 3

MARKS   STUDENT MARGIN

b) After compression, fuel is injected into the compressed air.

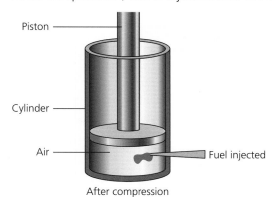

Piston

Cylinder

Air — — Fuel injected

After compression

The fuel ignites, causing combustion. This causes the pressure inside the cylinder to increase to 7.0×10^6 Pa.
The pressure of air outside the cylinder is 1.0×10^5 Pa.
The surface area of the piston is 6.4×10^{-3} m².

(i) Show that the force exerted on the piston is 4.4×10^4 N.

Space for working and answer

3   Properties of matter 3

(ii) This force causes the piston to be pushed upwards, resulting in an increase in the volume of the air in the cylinder.

Explain in terms of the kinetic model of gases why the pressure of the air inside the cylinder is reduced as the piston moves upwards.

1   Properties of matter 3

11 In a leisure swimming pool a wave machine produces water waves that travel across the pool. The waves are viewed from above.

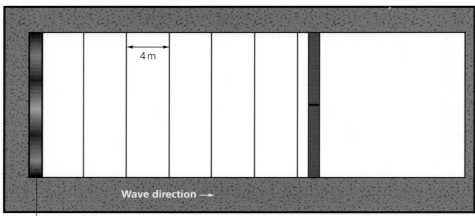

During one operation of the machine, waves of wavelength 4 m are produced at a frequency of 0·25 Hz.

a) **(i)** Calculate the period of the waves.

Space for working and answer

3   Waves 1

(ii) Calculate the speed of the waves.

Space for working and answer

3   Waves 1

| | MARKS | STUDENT MARGIN |

(iii) The swimming pool has barriers that can be opened to extend the pool. The barriers are opened to leave a 2m gap.

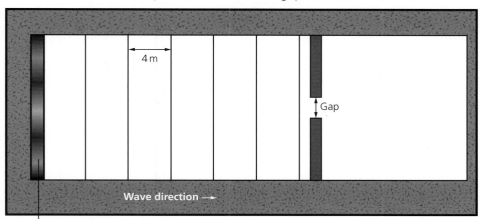

Complete the diagram to show the wave pattern on the right-hand side of the gap.

| | 2 | Waves 1 |

b) At a swimming competition, an underwater camera is used to view the competitors as they swim.

The underwater camera films a swimmer standing on the box outside the pool, waiting to start the race.

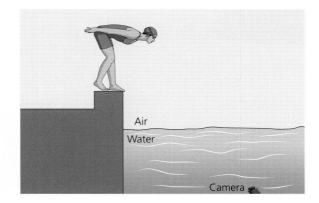

Complete the diagram to show the path of a ray of light from the swimmer to the camera.
You should include a normal in your diagram.

| | 3 | Waves 3 |

12 a) The diagram shows the electromagnetic spectrum.

Gamma rays	X-rays	Ultraviolet rays	Visible light		Microwaves	Radio waves

(i) Name the missing radiation.

| | 1 | Waves 2 |

(ii) State a detector of X-rays.

| | 1 | Waves 3 |

(iii) Which radiation band has the highest frequency?

| | 1 | Waves 3 |

b) Scientists used telescopes to detect microwaves from the Andromeda galaxy.

The galaxy was estimated to be 2.4×10^{22} m from Earth.

Determine this distance in light years.

Space for working and answer

3   Space 2

c) Before satellite communication was common, commercial shipping around the world received weather forecasts from land-based radio transmitters. Radio signals were broadcast on extremely long wavelengths.

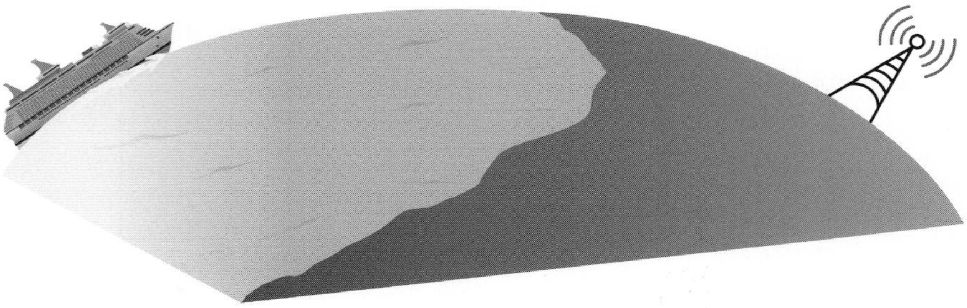

(i) Suggest why such long wavelengths were used to broadcast the shipping weather forecasts.

1   Waves 1

(ii) One such radio transmitter broadcast used a radio signal of wavelength 1500 m. The signal took 0.0046 s to reach a ship on the ocean.

Calculate the distance of the ship from the transmitter.

Space for working and answer

3   Waves 1

	MARKS	STUDENT MARGIN

13 Knowledge of the properties of waves in the electromagnetic spectrum has led to the development of many medical procedures for diagnosing and treating illnesses.

Use your knowledge of physics to comment on some of the developments. Make reference to typical sources, detectors and applications in your answer.

3 — Waves 2

14 A technician conducted an experiment to determine the half-life of a radioactive source. The apparatus used is shown.

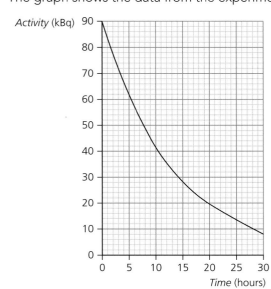

The source and detector were placed in a lead box.

a) Suggest why the experiment was carried out inside a lead box.

1 — Radiation 1

b) The graph shows the data from the experiment.

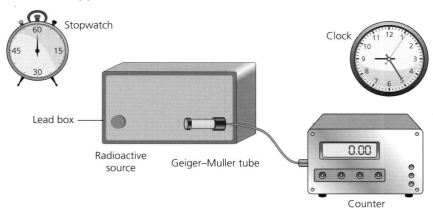

| | | MARKS | STUDENT MARGIN |

(i) What is meant by the term *half-life*?

1   Radiation 1

(ii) Use information from the graph to determine the half-life.

1   Radiation 1

c) A radioisotope was delivered to a hospital for use as a tracer in patients.

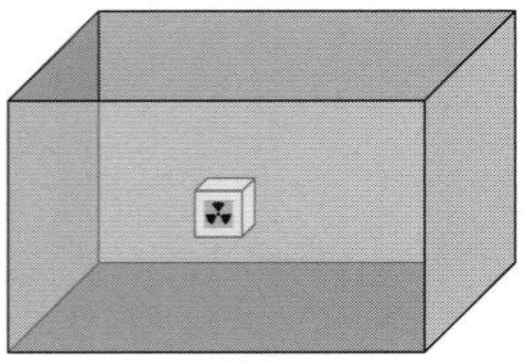

(i) While being transported to the hospital, the small box containing the radioisotope was positioned in the middle of a very large container.
Suggest a reason for this.
You must explain your answer.

2   Radiation 1

(ii) The activity of the radioisotope on delivery was 800 kBq at 7 a.m. on 5 July.
The half-life of the radioisotope is 8 hours.
The advice stated that the radioisotope should not be used if the activity has fallen below 100 kBq.
A radiologist wanted to use the radioisotope at 11 a.m. on 6 July.
Determine whether the radiologist should use the radioisotope.
Space for working and answer

2   Radiation 1

[END OF PRACTICE PAPER 1]

Section 1

Total marks: 25

Attempt ALL questions.

The answer to each question is **either** A, B, C, D or E. Decide what your answer is, then circle the appropriate letter. There is only **one correct answer** to each question.

Reference may be made to the Data sheet and to the Relationships sheet.

Allow yourself around 30 minutes for Section 1.

	MARKS	STUDENT MARGIN
	1	Dynamics 1

1 Four tugboats are used to give a resultant force on a cruise ship as it enters a harbour.

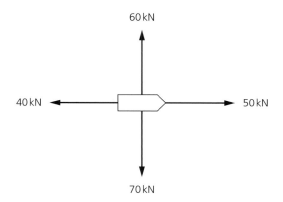

60 kN

40 kN

50 kN

70 kN

The forces applied by each boat are shown.

Which of the following forces could represent the resultant force on the cruise ship?

A

B

C

D

E

2 The graph shows how the velocity of an object changes with time.

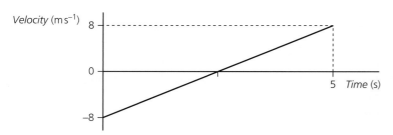

Which row in the table shows the displacement and acceleration of the object during the first 5 s?

	Displacement (m)	Acceleration (m s^{-2})
A	20	9·8
B	0	−9·8
C	20	32
D	0	−3·2
E	0	3·2

1   Dynamics 2

3 A student set up the apparatus as shown.

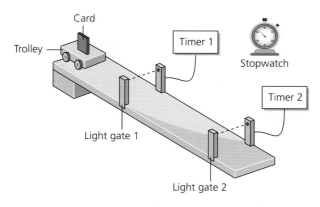

The trolley is released from rest and moves down the ramp.
The following measurements are recorded.

▶ Length of card = 0·06 m

▶ Time for card to pass through light gate 1 = 0·15 s

▶ Time for card to pass through light gate 2 = 0·04 s

▶ Time for trolley to pass between light gate 1 and light gate 2 = 0·50 s.
  The acceleration of the trolley between the light gates is

   A 0·45 m s^{-2}

   B 2·20 m s^{-2}

   C 3·60 m s^{-2}

   D 3·80 m s^{-2}

   E 6·40 m s^{-2}.

1   Dynamics 3

4 The velocity–time graph of an aircraft is shown as it lands on a runway.

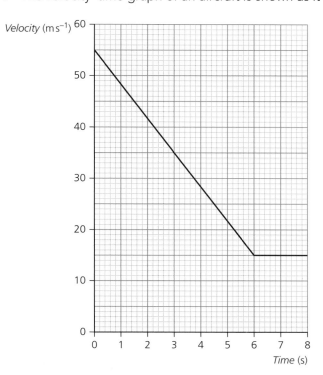

The acceleration of the aircraft during landing is

A $6{\cdot}7\,\mathrm{m\,s^{-2}}$

B $10{\cdot}0\,\mathrm{m\,s^{-2}}$

C $11{\cdot}7\,\mathrm{m\,s^{-2}}$

D $-6{\cdot}7\,\mathrm{m\,s^{-2}}$

E $-11{\cdot}7\,\mathrm{m\,s^{-2}}$.

MARKS   STUDENT MARGIN

5 Two forces act on a 4 kg metal block as shown.

1   Dynamics 3

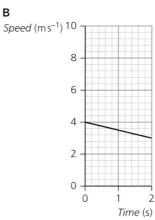

The block is initially moving at $4\,\text{m s}^{-1}$ in the direction shown.
The speed–time graph for the block is

A
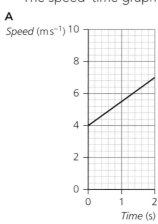

B
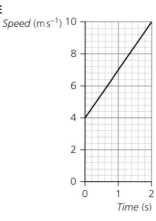

C
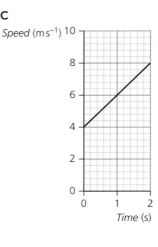

D
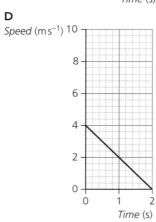

E

6 The graph shows how the gravitational field strength varies with height above the Earth's surface.

1   Dynamics 4

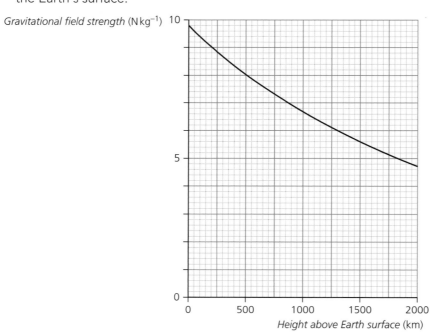

MARKS   STUDENT MARGIN

The International Space Station (ISS) orbits at 350 km above the Earth's surface.
A British astronaut on board the ISS in orbit has a mass of 90 kg.
The weight of the astronaut while on board the ISS is

A   0 N

B   90 N

C   765 N

D   828 N

E   882 N.

7 A ball is released from rest and allowed to roll 1·2 m down a straight track as shown.

1   Dynamics 5

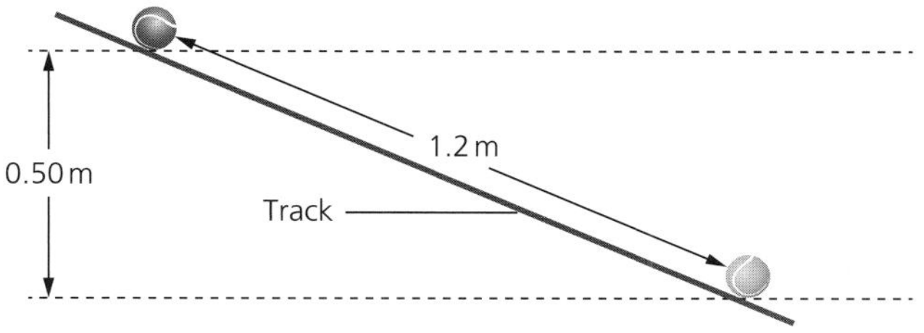

The mass of the ball is 0·40 kg. There is a constant frictional force of 8 N acting on the ball.
The change in the gravitational potential energy of the ball is

A   1·96 J

B   5·88 J

C   10·04 J

D   12·0 J

E   13·96 J.

8 In an experiment, two identical balls P and Q are projected horizontally from the edge of a table. The paths of the balls are shown in the diagram.

1   Dynamics 6

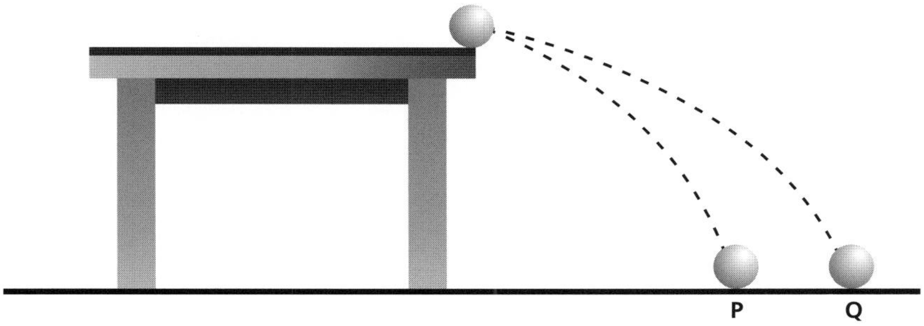

A student makes the following statements about the motion of the balls.

I   The balls have different horizontal velocities.

II   The balls take the same time to reach the floor.

III The balls have different final vertical velocities.

Which of the statements is/are correct?

A   I only

B   II only

C   III only

D   I and II only

E   II and III only.

9 Spacecraft are sometimes sent to explore distant planets or asteroids.

They have to travel large distances in space and can take many years to reach their destination. To increase their speed towards the destination planet, these spacecraft sometimes fly close by a planet during their journey in order to obtain a 'catapult' or 'gravity assist'.

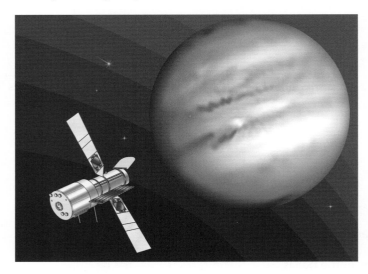

The speed of the spacecraft is increased when it leaves the planet because

A some of the spacecraft's kinetic energy is transferred to the planet

B there is no transfer of kinetic energy between the planet and the spacecraft

C the kinetic energy of the planet and the spacecraft increases

D some of the planet's kinetic energy is transferred to the spacecraft

E the kinetic energy of the planet and the spacecraft decreases.

1   Space 1

10 The Apollo 11 astronauts left the Moon in a space vehicle called the ascent module.

The ascent module had a mass of 5000 kg. At lift-off, the engine force was 15 000 N.

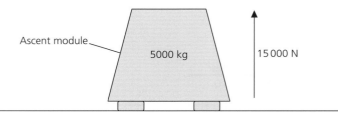

Ascent module

5000 kg   15 000 N

The acceleration of the module at lift-off was

A $1.4\,\text{m s}^{-2}$

B $2.0\,\text{m s}^{-2}$

C $3.0\,\text{m s}^{-2}$

D $4.6\,\text{m s}^{-2}$

E $6.8\,\text{m s}^{-2}$.

1   Space 1

	MARKS	STUDENT MARGIN

11 An ascending space rocket's engines exert a downward force on exhaust gases when in flight.

Which of the following is the reaction to this force?

A The exhaust gases exert an upward force on the space rocket's engines.
B The Earth exerts an upward force on the space rocket's engines.
C The air in the atmosphere exerts an upward force on the space rocket's engines.
D The exhaust gases exert a downward force on the air in the atmosphere.
E The Earth exerts an upward force on the exhaust gases.

MARKS: 1 — *STUDENT MARGIN:* Space 1

12 During early space exploration, Russia landed a lunar rover vehicle, named Lunokhod 2, on the Moon.

The vehicle carried out some basic research into the Moon's environment.

The weight of Lunokhod 2 on the Moon was 1360 N.

Which row in the table gives the mass and weight of Lunokhod 2 on Earth?

	Mass (kg)	Weight (N)
A	139	222
B	850	850
C	850	8330
D	1360	1360
E	2176	21 325

MARKS: 1 — *STUDENT MARGIN:* Space 1

MARKS | STUDENT MARGIN

13 A line spectrum obtained from a distant star is shown below.

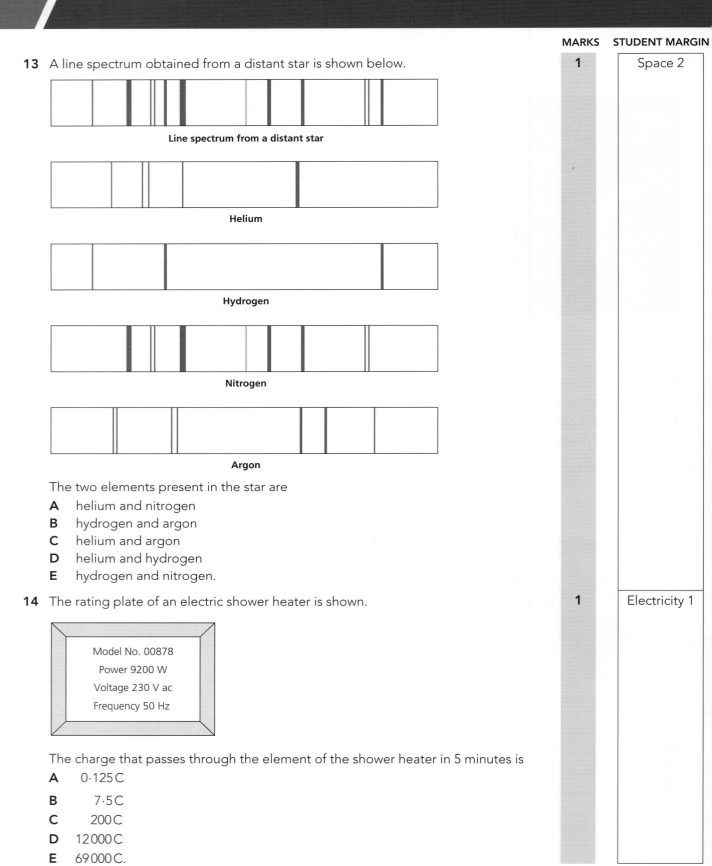

Line spectrum from a distant star

Helium

Hydrogen

Nitrogen

Argon

1   Space 2

The two elements present in the star are

A   helium and nitrogen

B   hydrogen and argon

C   helium and argon

D   helium and hydrogen

E   hydrogen and nitrogen.

14 The rating plate of an electric shower heater is shown.

1   Electricity 1

Model No. 00878
Power 9200 W
Voltage 230 V ac
Frequency 50 Hz

The charge that passes through the element of the shower heater in 5 minutes is

A   0·125 C

B   7·5 C

C   200 C

D   12000 C

E   69000 C.

MARKS	STUDENT MARGIN
1	Electricity 3

15 The circuit shows two resistors connected to a supply of 6V.

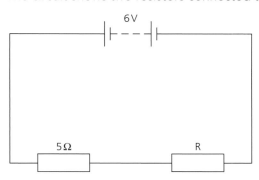

The current in the 5 Ω resistor is 0·5 A.

The resistance of resistor R is

A 5 Ω

B 7 Ω

C 12 Ω

D 17 Ω

E 60 Ω.

16 A student wishes to determine the value of an unknown resistor, R.

Which of the following circuits could provide readings to allow the use of Ohm's law to correctly determine a value for the resistance of R?

A
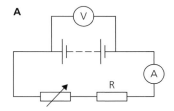

B
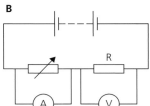

C
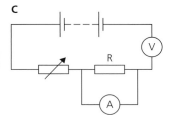

D
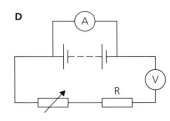

E
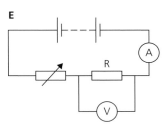

MARKS	STUDENT MARGIN
1	Electricity 3

17 Which of the following is the correct symbol for a light-emitting diode?

A   **B**   **C**   **D**   **E**

MARKS	STUDENT MARGIN
1	Electricity 4

	MARKS	STUDENT MARGIN

18 Four resistors are connected as shown.

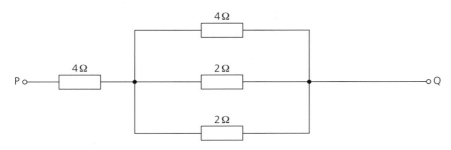

The resistance between P and Q is

A   0·67 Ω

B   3·25 Ω

C   4·8 Ω

D   5·25 Ω

E   7 Ω.

MARKS: **1**   STUDENT MARGIN: Electricity 4

19 Resistors are connected in the following circuit as shown.

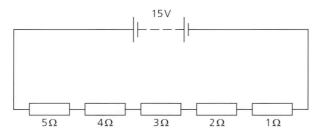

In which resistor is the greatest power developed?

A   1 Ω

B   2 Ω

C   3 Ω

D   4 Ω

E   5 Ω.

MARKS: **1**   STUDENT MARGIN: Electricity 5

20 Water at a temperature of 70 °C is cooled until it becomes ice at −15 °C.
The temperature change is

A   55 K

B   85 K

C   328 K

D   358 K

E   631 K.

MARKS: **1**   STUDENT MARGIN: Properties of matter 3

21 A syringe filled with air is sealed with a stopper.

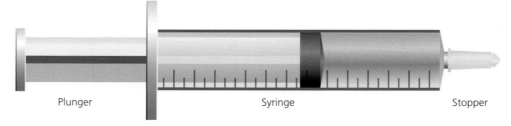

Plunger   Syringe   Stopper

MARKS: **1**   STUDENT MARGIN: Properties of matter 3

MARKS   STUDENT MARGIN

The plunger is slowly pressed into the syringe causing the volume of air to be reduced and the air pressure to increase. The temperature remains constant during this process.

Which of the following statements is/are correct?

I   The air molecules collide with the walls inside the syringe with greater force.

II   The air molecules collide with the walls inside the syringe more often.

III   The average speed of the air molecules inside the syringe increases.

A   I only

B   I and II only

C   II only

D   III only

E   I and III only.

22   A parakeet can hear sounds of wavelength ranging from 0·04 m to 1·7 m.

If the speed of sound in air is 340 m s^{-1}, the highest frequency the parakeet can hear is

A   13·6 Hz

B   200 Hz

C   578 Hz

D   8·5 kHz

E   85 kHz.

1   Waves 1

23   For a ray of light travelling from air into glass, which of the following statements is/are correct?

I   The speed of light sometimes changes.

II   The speed of light always changes.

III   The wavelength of light sometimes changes.

IV   The wavelength of light always changes.

A   I only

B   III only

C   I and III only

D   II and III only

E   II and IV only.

1   Waves 3

24 A ray of green light emerges from a glass block into air.

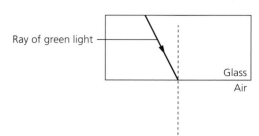

Which diagram correctly represents the ray of light when it emerges from the glass block?

A

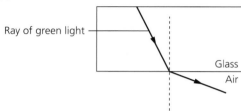

B

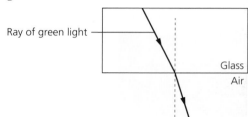

C

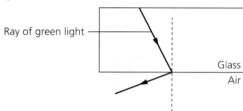

D

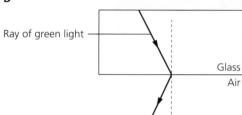

E

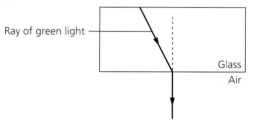

25 A worker in the nuclear industry received an equivalent dose of $0.08\,\mu$Sv in 16 hours.

The equivalent dose rate is

A $1.39 \times 10^{-12}\,\text{Sv}\,\text{h}^{-1}$

B $5.0 \times 10^{-9}\,\text{Sv}\,\text{h}^{-1}$

C $1.61 \times 10^{-7}\,\text{Sv}\,\text{h}^{-1}$

D $1.28 \times 10^{-6}\,\text{Sv}\,\text{h}^{-1}$

E $4.61 \times 10^{-3}\,\text{Sv}\,\text{h}^{-1}$.

Section 2

Total marks: 110

Attempt ALL questions.

Write your answers clearly in the spaces provided. If you need additional space for answers or rough work, please use separate pieces of paper.

Reference may be made to the Data sheet and to the Relationships sheet.

Allow yourself around 2 hours for Section 2.

	MARKS	STUDENT MARGIN

1 A technician sets up an experiment to investigate the effect of an electric field on charged particles. An electric field is produced between two metal plates inside a container, at the top and bottom, and connected to a power supply.

Charged particles are placed inside the container and their movement recorded.

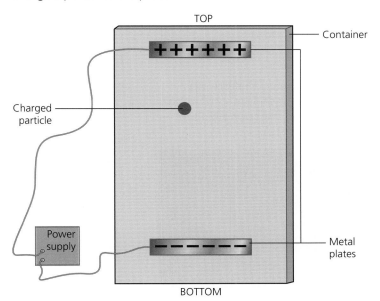

During one test, a charged particle is placed in the container.

The forces acting on the particle are shown in the diagram below.

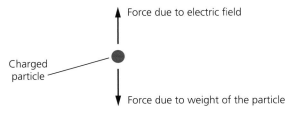

The particle remains stationary inside the electric field.

a) State the sign of the charge on the particle.

MARKS: 1 — Electricity 2

b) The mass of the particle is $5 \cdot 4 \times 10^{-8}$ kg.

Calculate the weight of the particle.

Space for working and answer

MARKS: 3 — Dynamics 4

MARKS STUDENT MARGIN

c) The electric field is switched off and the particle begins to fall through the air inside the container.

A velocity–time graph of the downward motion of the facing particle is shown.

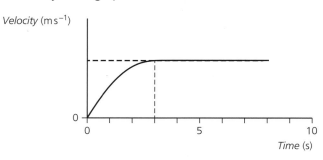

Describe and explain the movement of the particle after 3 s.

2   Dynamics 4

2 While discussing different car journeys, a driving enthusiast says that driving 20 km along a motorway at high speed within the speed limit of $31\,\text{m s}^{-1}$ (70 mph) probably requires less fuel than driving 20 km in town within the speed limit of $13\,\text{m s}^{-1}$ (30 mph).

Use your knowledge of physics to comment on whether motorway or town driving may require less fuel.

3   Dynamics 5

MARKS	STUDENT MARGIN

3 A student is a keen archer. The student investigates the average force required to stretch the string of the bow before firing an arrow.

The student attaches a force-sensor to the bow string to measure the average force required to stretch the string by different amounts.

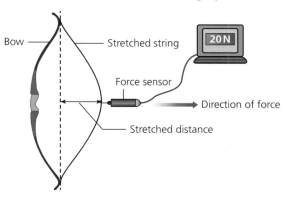

Bow — Stretched string — 20 N

Force sensor

→ Direction of force

Stretched distance

The student plots a graph of the average force applied to the string and its stretched distance.

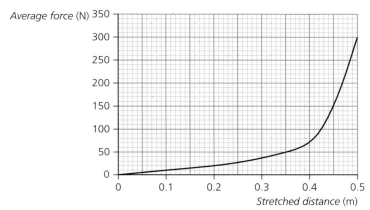

Average force (N)

Stretched distance (m)

a) **(i)** State the average force required to stretch the bow string by 0·35 m.

1   Dynamics 4

    (ii) Calculate the work done in stretching the string by 0·35 m.
        Space for working and answer

3   Dynamics 5

b) During a practice session, the student releases an arrow of mass 0·06 kg horizontal towards a target. The distance to the target is 30 m.

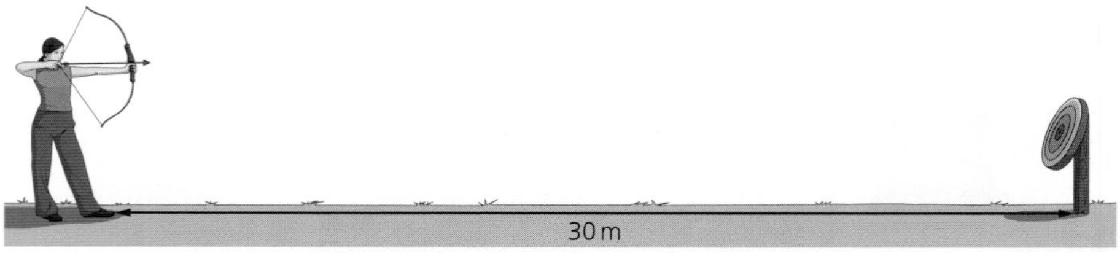

30 m

(i) The arrow leaves the bow with a horizontal velocity of 40 m s⁻¹.
Calculate the time of flight of the arrow until it reaches the target.
The effects of air resistance can be ignored.
Space for working and answer

3 Dynamics 6

(ii) Calculate the kinetic energy of the arrow as it leaves the bow.
Space for working and answer

3 Dynamics 5

(iii) The vertical velocity of the arrow as it reaches the target is 7·35 m s⁻¹.
By scale diagram, or otherwise, determine:
(A) the magnitude of the resultant velocity of the arrow as it reaches the target;
Space for working and answer

2 Dynamics 1

(B) the direction of the resultant velocity of the arrow as it reaches the target.
Space for working and answer

2 Dynamics 1

MARKS   STUDENT MARGIN

4   Read the passage below and answer the questions that follow.

Black holes

A black hole is where the force of gravity is so strong because matter has collapsed into a tiny space. This can happen when a star has used up most of its energy.

Because no light can get out, black holes cannot be seen. Space telescopes can help find black holes. Using special instruments on satellites, the behaviour of stars that are close to black holes can be monitored.

Black holes can have a range of sizes. A 'stellar' black hole has a mass of up to 20 times more than the mass of the Sun. There may be many stellar mass black holes in the Milky Way galaxy.

There is evidence which suggests that at the centre of the Milky Way there is a supermassive black hole.

A supermassive black hole has a mass of more than 1 million suns. Scientists have discovered that every large galaxy contains a supermassive black hole at its centre.

At a distance of 27 600 light years from Earth, the supermassive black hole at the centre of the Milky Way is called Sagittarius A. Sagittarius A would fit inside a large sphere which could hold several million Earth masses.

Scientists think the supermassive black holes formed when the universe began.

Stellar black holes are made when the centre of a very big star collapses inwards on itself. When this happens, it causes a supernova. A supernova is an exploding star that blasts part of the star into space.

Black holes cannot be seen directly because gravity prevents light escaping from the black hole. Observation of how nearby stars are affected by their strong gravity provides information about the behaviour, size and nature of the black hole.

The interaction of stars and black holes when they are close together produces intense gamma radiation. Satellites and telescopes in space are used to detect this radiation.

a)   Name a black hole mentioned in the passage.   1   Space 1

b)   Approximately how many years ago do scientists believe that the supermassive black holes were formed?   1   Space 2

c)   Calculate the distance of the Earth, in metres, from the centre of the Milky Way.   3   Space 2

Space for working and answer

d)   Telescopes on satellites are used to detect light rays and gamma radiation. Name a detector of gamma rays.   1   Waves 2

5   The table lists three telescopes on satellites in orbit around the Earth that are used to monitor different bands of the electromagnetic spectrum of radiation detected from space.

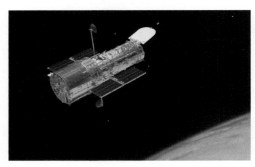

Name of satellite	Band of electromagnetic spectrum monitored	Approximate altitude above Earth (km)
WISE	infrared	525
Chandra	X-rays	105 000
IRIS	ultraviolet	650

a) (i) List the satellites in order of increasing wavelength of the electromagnetic spectrum band monitored.

2   Waves 2

1. _____

2. _____

3. _____

(ii) List the satellites in order of increasing period of orbit around the Earth.

2   Space 1

1. _____

2. _____

3. _____

b) While monitoring signals from space, the IRIS satellite detects electromagnetic radiation with a wavelength of $2·5 \times 10^{-7}$ m.

Calculate the frequency of the radiation.

3   Waves 1

Space for working and answer

c) The Chandra satellite sends a radio signal to Earth.

Calculate the time for the radio signal to reach Earth.

3   Dynamics 1

Space for working and answer

MARKS | STUDENT MARGIN

d) Explain why telescopes on satellites in orbit around the Earth are required to monitor some electromagnetic radiation from space, in addition to Earth-based telescopes.

1 | Space 1

6 A student investigates the resistance of a lamp. The student uses the circuit shown to measure the current in the lamp at different voltages.

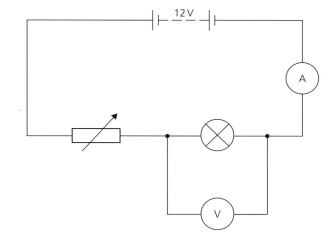

a) Explain how the current can be altered in this circuit.

1 | Electricity 4

b) The student recorded the voltage across the lamp for different values of current in the lamp.
The table shows the student's results.

Voltage (V)	Current (A)
0	0
0·4	0·18
1·6	0·44
2·8	0·60
4·4	0·76
6·4	0·90
9·6	1·00

MARKS STUDENT MARGIN

(i) Using the graph paper, draw a graph of these results.

3 Electricity 3

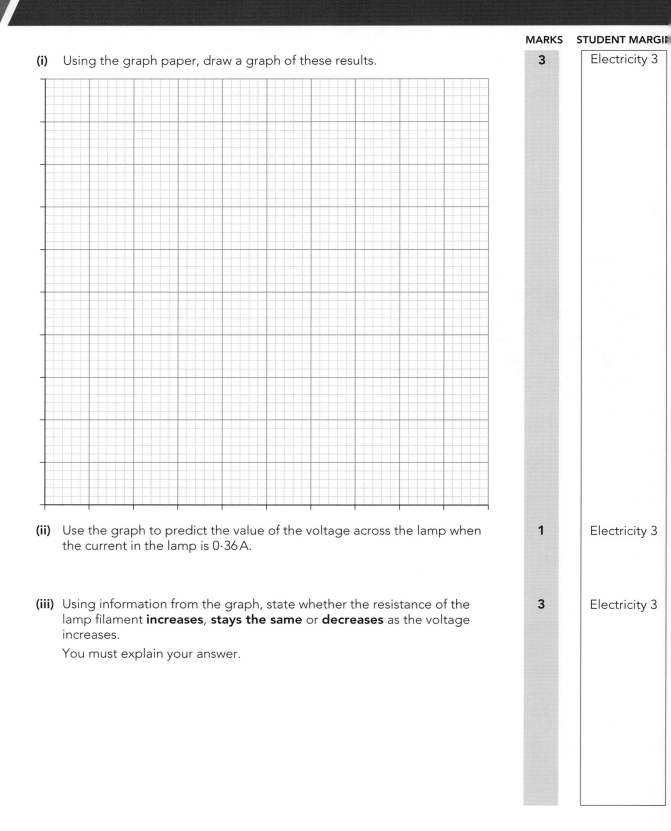

(ii) Use the graph to predict the value of the voltage across the lamp when the current in the lamp is 0·36 A.

1 Electricity 3

(iii) Using information from the graph, state whether the resistance of the lamp filament **increases**, **stays the same** or **decreases** as the voltage increases.

You must explain your answer.

3 Electricity 3

	MARKS	STUDENT MARGIN

7 Three new ceiling lamps are installed in a kitchen.

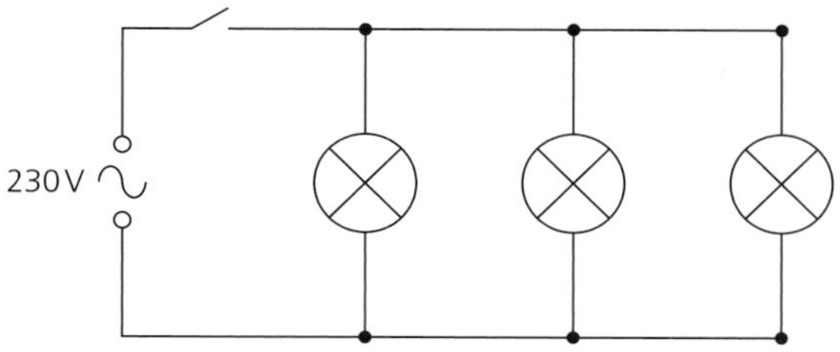

The lamps are connected to the mains in the circuit shown.

Each lamp is rated at 230V, 50W.

a) **(i)** Explain why the lamps must be connected in parallel.

1   Electricity 4

(ii) Calculate the resistance of each lamp when operating at their correct voltage.

Space for working and answer

3   Electricity 5

b) Calculate the total current when all three lamps are operating.

Space for working and answer

3   Electricity 5

MARKS   STUDENT MARGIN

8   A student cooks vegetables using an electric steamer. Water in the base of the steamer is heated, causing steam to rise through holes in the base of two vegetable compartments. The steam cooks the vegetables.

Steam

Vegetables

Water

The steamer has a rating of 800 W and is filled with 0·6 kg of water at a temperature of 25 °C.

a)   Calculate the heat energy required to heat the water from 25 °C to its boiling point of 100 °C.

Space for working and answer

3

Properties of matter 1

b)   The vegetables need to be steamed for 12 minutes after the water has reached its boiling point of 100 °C.

(i)   Calculate the electrical energy supplied to the steamer during this time.

Space for working and answer

3

Electricity 5

(ii)   Calculate the maximum mass of water that could be converted into steam in this time.

Space for working and answer

3

Properties of matter 2

	MARKS	STUDENT MARGIN

(iii) Explain why the mass calculated in part **b) (ii)** is the maximum mass of steam that can be produced.

1   Dynamics 5

9 a) State the definition of *pressure*.

1   Properties of matter 3

b) To remain safe when diving, deep sea divers must know the pressure exerted on them by the sea at different depths.

The pressure exerted on deep sea divers when diving beneath the sea is calculated using the relationship:

$p = \rho g h$

where:

- p is the pressure exerted by the sea in Pa
- ρ is the density of the sea water in $kg\,m^{-3}$
- g is the acceleration due to gravity in $m\,s^{-2}$
- h is the submerged depth of the diver in metres.

During one dive, a diver reaches a depth of 24 m. The density of the water is $1025\,kg\,m^{-3}$.

Calculate the pressure exerted by the sea water on the diver at this depth.

Space for working and answer

3   Properties of matter 3

10 Scientists at a research station located in the Antarctic use a steel drum to store equipment.

The volume of the container is $0.25\,m^3$. The container has a lid that forms an airtight seal when attached to the container.

One empty container is left outside with the lid secured in position.

The pressure of the air inside the container is 1.01×10^5 Pa. The temperature of the air inside the container is 10 °C.

During the night, the temperature of the air inside the container falls to −40 °C.

a)   Calculate the air pressure inside the container at −40 °C.

Assume that the volume of the container remains constant.

Space for working and answer

3   Properties of matter 3

b)   The air pressure outside the container remains constant at 1.01×10^5 Pa.

(i)   Calculate the pressure difference between the air outside and inside the container at −40 °C.

Space for working and answer

1   Properties of matter 3

(ii) The surface area of the lid is $0.45\,m^2$.
Calculate the force on the lid due to the pressure difference.

Space for working and answer

3   Properties of matter 3

(iii) State the direction of this force on the lid.

1   Dynamics 1

	MARKS	STUDENT MARGIN

11 a) Explain the difference between alternating and direct current.

1   Electricity 1

b) A student uses an oscilloscope, resistor and signal generator to investigate alternating current.

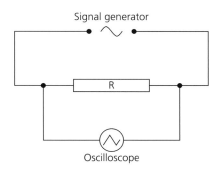

Signal generator

R

Oscilloscope

The signal generator produces an alternating voltage signal. The voltage and frequency of the signal can be varied.

The oscilloscope has controls which can be adjusted to allow the amplitude and period of an electrical voltage signal to be measured.

The following oscilloscope display was produced for one particular setting of voltage and frequency of the signal generator.

The voltage and time scale settings for the oscilloscope are shown. The oscilloscope grid is made up of 1 cm squares.

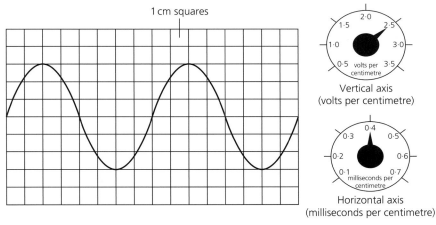

1 cm squares

Vertical axis
(volts per centimetre)

Horizontal axis
(milliseconds per centimetre)

(i) What is the amplitude of the voltage supply?

1   Waves 1

(ii) What is the period of the wave?

1   Waves 1

(iii) Calculate the frequency of the wave.
Space for working and answer

3   Waves 1

c) The student constructs the following circuits using light-emitting diodes (LEDs) connected to d.c. and a.c. power supplies.

State which of the LEDs W, X, Y or Z will light:

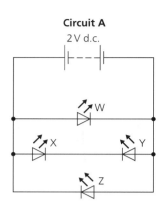

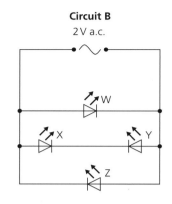

(i) in circuit A;

1   Electricity 4

(ii) in circuit B.

1   Electricity 4

12 A student takes a dog for a walk at night in a field. The student can hear the dog barking in the distance ahead, but cannot see the dog in the beam of the torch.

Use your knowledge of physics to comment on why the student can hear the dog but cannot see the dog.

3   Waves 1

13 Smoke detectors are fitted in houses to sound an alarm when smoke is detected.

The smoke detector circuit is normally housed inside a plastic container.

The detector uses a radioactive source that emits alpha radiation.

The detector contains two metal electrodes that are connected to a power supply. The gap between the electrodes contains air.

The radioactive source is placed beside this gap. Radiation from the source ionises the air between the electrodes and causes a current in the circuit.

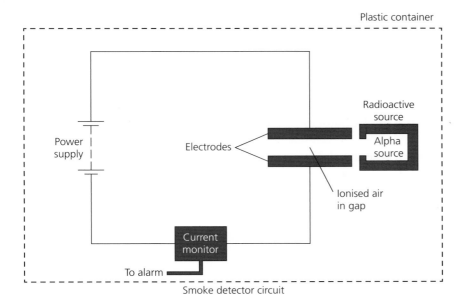

When smoke from a fire enters the gap between the electrodes, this reduces the current. A current monitor detects this change in current and sets off an alarm.

a) State what is meant by the term *alpha radiation*.

1   Radiation 1

b) Explain what is meant by *ionisation*.

1   Radiation 1

c) Suggest **two** reasons why a radioactive source that emits alpha radiation is preferred for this application, instead of a source which emits beta or gamma radiation.

2   Radiation 1

	MARKS	STUDENT MARGIN

d) State the current annual effective dose exposure safety limit of radiation for a member of the public.

1 — Radiation 1

e) The table shows a list of radioactive sources which are available for use by a manufacturer of smoke detectors. The radioactive sources all emit alpha radiation.

Radioactive source	Half-life
radium-223	11 days
polonium-210	138 days
americium-241	432 years

State which source the manufacturer should select.
Explain your answer.

2 — Radiation 1

f) For the radioactive source used in the smoke detector, 1.68×10^7 atoms decay in 7 minutes.
Determine the activity of the radioactive source.
Space for working and answer

3 — Radiation 1

14 Currently, nuclear power stations use nuclear fission as the source of energy for the production of electricity. During the fission process, an atomic particle is absorbed by a uranium nucleus. Fission products and neutrons are produced.

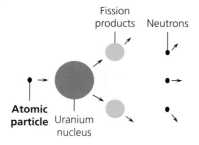

Fission
products   Neutrons

**Atomic
particle** Uranium
nucleus

a) Name the atomic particle absorbed by the uranium nucleus during nuclear fission.

1 — Radiation 1

MARKS   STUDENT MARGIN

b) During research of the fission process, a graph was produced showing the activity against time for a sample of one of the fission products.

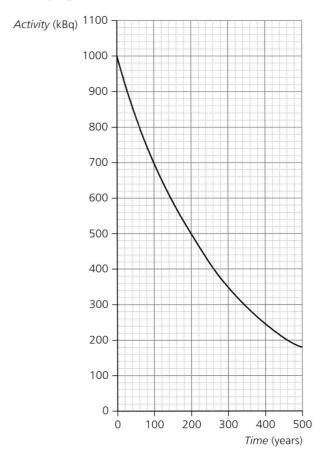

Activity (kBq)

Time (years)

(i) Determine the half-life of the fission product.

1   Radiation 1

(ii) The safe level of activity for this sample is 62·5 kBq.
Calculate the time taken for the activity of this sample to fall to this level.

2   Radiation 1

Space for working and answer

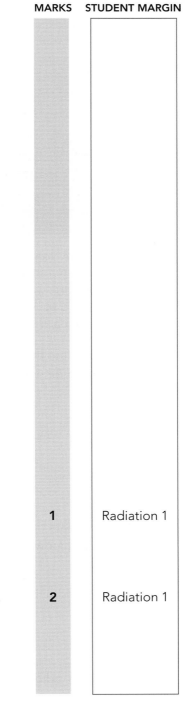

c) Some nuclear power stations have reached the end of their predicted lifetime of operation and must be dismantled.

During the dismantling process workers are regularly monitored for exposure to radiation.

A demolition worker of mass 70 kg was found to have been exposed to a radioactive source emitting alpha particles.

Across the whole body, the worker absorbed 850 μJ of energy.

(i) Calculate the absorbed dose received by the worker.

Space for working and answer

3   Radiation 1

(ii) Calculate the equivalent dose received by the worker.

Space for working and answer

3   Radiation 1

(iii) State one precaution that the worker could observe to reduce exposure to the radioactive source.

1   Radiation 1

d) Scientists are researching another source of nuclear energy as an alternative to nuclear fission.

State the name of this nuclear process.

1   Radiation 1

[END OF PRACTICE PAPER 2]

ANSWERS TO PRACTICE PAPERS

Practice Paper 1

Section 1

Question	Answer	Max. mark	Commentary with hints and tips
1	C	1	Hint: it is sometimes helpful to mark each quantity in the answers with 'v' for a vector and 's' for a scalar to make the correct selection of the answer easier.
2	D	1	For average speed, use $d = vt$, $(50 + 30) = v \times 8$, $v = 10\,\text{ms}^{-1}$. For average velocity, use $s = \bar{v}t$, $(50 - 30) = \bar{v} \times 8$, $\bar{v} = 2\cdot5\,\text{ms}^{-1}$ South.
3	A	1	Use acceleration $a = \dfrac{v - u}{t} = \dfrac{4 - 0}{5} = 0\cdot8\,\text{ms}^{-2}$.
4	E	1	Use $W = mg = 1 \times 9\cdot8 = 9\cdot8\,\text{N}$ to calculate the weight of the ball. The weight of the ball is the downward force of gravity acting on the ball at all times.
5	C	1	Each ball takes the same time to reach sea level because they are released from the same height and have the same vertical acceleration. Air resistance is ignored in this question, so there is no horizontal force acting on the balls; there is no horizontal acceleration. Different horizontal distances travelled mean that the horizontal velocities are different because the balls are in the air for the same time.
6	D	1	You need to be able to identify the terms *planet*, *dwarf planet*, *moon*, *Sun*, *asteroid*, *solar system*, *star*, *exoplanet*, *galaxy* and *universe* correctly and in context.
7	C	1	The period of a satellite increases as its orbital height increases. 20 000 km is greater than 1340 km and less than 35 900 km, so the period of a satellite in orbit at 20 000 km must be between the periods of satellites at these altitudes. The only one in the list is 720 minutes.
8	E	1	You must know the approximate estimated age of the universe, which is 14 billion years to the nearest billion.
9	D	1	The definition of electrical current is the electric charge transferred per unit time, which is one coulomb per second.
10	D	1	First calculate the current using Ohm's law: $I = \dfrac{V_s}{\text{total resistance}} = \dfrac{60}{30} = 2\,\text{A}$ The charge passing through the series circuit has the same value at all positions. The relationship used to calculate charge is $Q = It$ (convert 2 minutes into seconds).
11	D	1	You need to know the path of a charged particle between two oppositely charged parallel plates or near a single point charge or **between two oppositely charged points** or between two like charged points. Protons follow electric field lines from positive to negative, electrons from negative to positive.
12	B	1	Use Ohm's law, $V = IR$, to calculate each resistor using values for V and I from the graph. $V = IR_P$, $20 = 2 \times R_P$, $R_P = 10\,\Omega$ $V = IR_Q$, $10 = 4 \times R_Q$, $R_Q = 2\cdot5\,\Omega$
13	B	1	LDR: The circuit symbol, function and application of standard electrical and electronic components should be studied and memorised.

Question	Answer	Max. mark	Commentary with hints and tips
14	E	1	When the variable resistor's resistance is decreased, the share of the battery voltage across the variable resistor (voltmeter V_1) decreases and the share of the battery voltage across the fixed resistor (voltmeter V_2) increases. Also, the total resistance in the circuit decreases. This means that the current (and ammeter reading) increases.
15	D	1	The horizontal section of the graph between 100 and 350 seconds indicates where the substance is changing state from solid to liquid. The energy supplied to the substance during this time is calculated using $P = \dfrac{E}{t}$, $200 = \dfrac{E}{(350-100)}$, $E = 50\,000\,\text{J}$. To calculate specific latent heat of fusion of the substance use $E = ml_f$, $50\,000 = 0\cdot25 \times l_f$, $l_f = 2\cdot0 \times 10^5\,\text{J}\,\text{kg}^{-1}$.
16	B	1	The volume of a fixed mass of gas at constant temperature is inversely proportional to its pressure, so a graph of pressure p against $\dfrac{1}{V}$ will produce a straight-line graph through the origin.
17	C	1	Wavespeed $v = \dfrac{\text{distance travelled}}{\text{time}} = \dfrac{d}{t} = \dfrac{85}{2\cdot5} = 34\,\text{ms}^{-1}$ Frequency $f = \dfrac{N}{t} = \dfrac{5}{2\cdot5} = 2\,\text{Hz}$ Amplitude $a = \dfrac{30}{2} = 15\,\text{m}$
18	D	1	When light travels from air into glass: ▸ its speed decreases ▸ its frequency stays constant ▸ its wavelength decreases.
19	A	1	You need to know the nature of alpha (α), beta (β) and gamma (γ) radiation in terms of which particles alpha and beta consist of, and that gamma is electromagnetic radiation.
20	E	1	Both alpha and beta radiation are absorbed by 5 mm of aluminium, so only gamma radiation is detected.
21	C	1	Use $A = \dfrac{N}{t} = \dfrac{30\,000}{5\times60} = 100\,\text{Bq}$ (remember t must be converted into seconds).
22	A	1	Use $D = \dfrac{E}{m} = \dfrac{2\times10^{-3}}{0\cdot08} = 0\cdot025\,\text{Gy}$.
23	B	1	Use the Data sheet to obtain a value for the radiation weighting factor, w_r, for slow neutrons, which is 3, then use: $H = Dw_r = 50 \times 10^{-3} \times 3 = 0\cdot15\,\text{Sv}$
24	E	1	You need to know the equivalent dose rate and exposure safety limits for the public and for workers in radiation industries in terms of annual effective equivalent dose required. Currently, the annual exposure safety limits are: ▸ 1 mSv for the public ▸ 20 mSv for workers in radiation industries.
25	C	1	Only gamma radiation can penetrate through the skin to the detector, so the tracer must emit gamma radiation. The radioactive tracer should also have a short half-life to limit the presence of the radiation inside the body.

Section 2

Question			Answer	Max. mark	Commentary with hints and tips
1	a)	(i)	$a = \dfrac{v-u}{t}$ (1) $= \dfrac{80-0}{60}$ (1) $= 1\cdot3\,\mathrm{m\,s^{-2}}$ (1)	3	Care is required to select the correct values for u ($0\,\mathrm{m\,s^{-1}}$) and v ($80\,\mathrm{m\,s^{-1}}$) from the graph, and to use them correctly in $a = \dfrac{v-u}{t}$.
		(ii)	unbalanced force $F_{un} = ma$ (1) $= 360\,000 \times 1\cdot3$ $= 468\,000\,\mathrm{N}$ (1) average frictional force = forward force of engines $- F_{un}$ $= 500\,000 - 468\,000$ (1) average frictional force $= 32\,000\,\mathrm{N}$ (1)	4	First, calculate the unbalanced force required to produce the aircraft's acceleration using the value for acceleration calculated in part **a) (i)**. The forward force overcomes the average frictional force and provides the unbalanced accelerating force: forward force = average frictional force $+ F_{un}$
		(iii)	Total distance = area under graph $= \dfrac{1}{2}bh$ (1) $= \dfrac{1}{2} \times 60 \times 80$ (1) $= 2400\,\mathrm{m}$ (1)	3	The aircraft takes off after 60 seconds, so the total distance required during take-off is the area under the graph.
	b)		At constant height, upward force = weight of aircraft $= mg$ (1) $= 360\,000 \times 9\cdot8$ (1) $= 3\cdot53 \times 10^6\,\mathrm{N}$ (1)	3	When the aircraft's height is constant, the upward force on the aircraft and its weight are balanced.
	c)		Exposure time in 1 year = time for one flight × number of flights $= 7 \times 106$ $= 742$ hours $\dot{H} = \dfrac{H}{t}$ (1) $8 \times 10^{-6} = \dfrac{H}{742}$ (1) $H = 5\cdot9\,\mathrm{mSv}$ (1)	3	First, calculate the total number of hours of exposure. When writing the relationship, $\dot{H} = \dfrac{H}{t}$, it is important to correctly represent equivalent dose rate, $\dot{H}$, with a clearly marked dot above. Use time, t, in the same units as given in the question (hours).
				(16)	
2	a)		$E_k = \dfrac{1}{2}mv^2$ (1) $= \dfrac{1}{2} \times 500 \times 55^2$ (1) $= 7\cdot6 \times 10^5\,\mathrm{J}$ (1)	3	When using $E_k = \dfrac{1}{2}mv^2$ to calculate the kinetic energy, remember to multiply by the velocity squared.

Question	Answer	Max. mark	Commentary with hints and tips
b)	$E_k \equiv E_p$ (1) $7{\cdot}6 \times 10^5 = mgh$ (1) $7{\cdot}6 \times 10^5 = 500 \times 9{\cdot}8 \times h$ (1) $\quad\quad h = 155\,\text{m}$	3	When asked to show how a value is reached, take care to include important steps. First mark awarded for a statement that energy is conserved, i.e. that the kinetic energy of the water is transformed into gravitational potential energy. Second mark awarded for stating the relationship used to calculate E_p. Third mark awarded for showing how various quantities are used within the relationship $E_p = mgh$, which lead to the final answer.
c)	Frictional forces between the jet and air cause some kinetic energy to be lost and not converted into gravitational potential energy.	1	As the water rises through the air, air friction causes some kinetic energy to be changed into heat and some sound energy, reducing the energy available to be converted into potential energy.
		(7)	
3	**Sample answer** ▸ Stationary objects have balanced forces acting, e.g. when a person is sitting, their weight is balanced by an upward reaction force from the chair. ▸ When an unbalanced force acts on an object, it causes acceleration which is proportional to the size of the unbalanced force, e.g. when a vehicle accelerates from rest, an unbalanced force acts on it causing an acceleration which is proportional to the size of the unbalanced force acting on the vehicle and inversely proportional to the mass of the vehicle. ▸ There is an equal and opposite reaction force to every action force, e.g. when a person walks, their foot exerts a backward force on the ground, while the ground exerts a forward force on the foot, causing the person to move forward.	3	This is an open-ended question: a variety of physics statements and descriptions can be used to answer this question. Marks are awarded on the basis of whether the answer overall demonstrates 'no' (0 marks), 'limited' (1 mark), 'reasonable' (2 marks) or 'good' (3 marks) understanding. 3 marks would be awarded to an answer which demonstrates a good understanding of the physics involved. The answer would show a good comprehension of the physics of the situation, provided in a logically correct sequence. (This type of answer might include a statement of the principles involved, a relationship or an equation, and the application of these to respond to the problem.) Also, note that the open-ended type of question is worth 3 marks. It is important not to spend too much time answering this question (3–4 minutes is the average time in the paper for 3 marks). A guide would be to take perhaps 3–5 minutes to complete this type of question. Only use more time than this if you have completed and checked all of the remaining questions in the paper.
		(3)	

Question			Answer		Max. mark	Commentary with hints and tips
4	a)		Sunspots are caused by intense magnetic fields appearing beneath the Sun's surface.		1	Careful reading of the passage is required to match the relevant part of the text to the question being asked.
	b)		Ultraviolet radiation and X-rays		2	Look for keywords in the text which match the words in the question. Here, the mention of satellites should direct you to the section about satellite damage.
	c)		$d = vt$ $2.28 \times 10^{11} = v \times 2.3 \times 24 \times 60 \times 60$ $v = 1.15 \times 10^6\,\mathrm{m\,s^{-1}}$	(1) (1) (1)	3	Sometimes, as in this question, you are required to make a calculation using information from the text. Take care to use the correct values from the text, converting days into seconds, for example.
					(6)	
5	a)	(i)	Distance travelled $= 125 - 11 = 114\,\mathrm{km}$ $E_w = Fd$ $= 3.7 \times 10^5 \times 114 \times 10^3$ $= 4.2 \times 10^{10}\,\mathrm{J}$	(1) (1) (1)	3	Calculate the distance that the average frictional force acts across. Remember to convert km into m.
		(ii)	Friction between spacecraft and atmosphere produces heat energy which may cause damage to the contents of the spacecraft.		1	The contents (controls, apparatus, etc.) of the spacecraft require protection from extreme temperatures.
		(iii)	Moon has no atmosphere so there would be no frictional forces acting on the parachute or spacecraft to slow it down.		1	Upward frictional forces acting on the parachute decelerated the spacecraft in the Martian atmosphere.
	b)	(i)	$W = mg$ $12210 = m \times 3.7$ $m = 3300\,\mathrm{kg}$	(1) (1)	2	Use the Data sheet to obtain the value for g, the gravitational field strength on Mars ($3.7\,\mathrm{N\,kg^{-1}}$).
		(ii)	Upward unbalanced force $=$ $18810 - 12210 = 6600\,\mathrm{N}$ $F_{un} = ma$ $6600 = 3300 \times a$ $a = 2\,\mathrm{m\,s^{-2}}$	(1) (1) (1) (1)	4	First, calculate the unbalanced upward decelerating force. Since the question asks for the magnitude of the deceleration, it is not necessary to include a negative sign.
					(11)	
6	a)	(i)	2V		1	Care is required when extracting data from graphs; a ruler is useful when obtaining these values.
		(ii)	Voltage across the capacitor is changing, so the voltage across the resistor is changing and so the current in the resistor changes.		1	Current is the rate of movement of charge in a conductor. As the capacitor charges, this rate reduces so the current reduces in its circuit.
		(iii)	$V_R = 3 - 2 = 1\,\mathrm{V}$ $I = \dfrac{V}{R}$ $= \dfrac{1}{2000}$ $= 5 \times 10^{-4}\,\mathrm{A}$	(1) (1) (1) (1)	4	Use the voltage across the capacitor from part **a) (i)** to calculate the voltage across the resistor after 3 minutes. Use Ohm's law to determine the resistance of R.

Question			Answer		Max. mark	Commentary with hints and tips
	b)		The voltage across the capacitor increases and reaches a value which will switch the MOSFET on (to sound the buzzer)	(1) (1)	2	The voltage across the capacitor increases and eventually has a value which will switch on MOSFET, causing a current in the buzzer.
					(8)	
7	a)	(i)	$I = \dfrac{P}{V}$ $I = \dfrac{18}{12}$ $= 1 \cdot 5\,\text{A}$	(1) (1) (1)	3	At correct power rating, the voltage across each lamp is 12V, the power rating is 18W. The supply voltage should not be used here.
		(ii)	Reading on voltmeter = supply voltage − voltage across each lamp $= 30 - (2 \times 12)$ $= 6\,\text{V}$	(1)	1	The voltmeter is placed across the variable resistor. The lamps are in series with the variable resistor, so the voltages across each component add up to 30V.
	b)	(i)	$\dfrac{1}{R_T} = \dfrac{1}{R_1} + \dfrac{1}{R_2} + \dfrac{1}{R_3}$ $= \dfrac{1}{8} + \dfrac{1}{8} + \dfrac{1}{16}$ $R_T = 3 \cdot 2\,\Omega$	(1) (1) (1)	3	Take care to insert the correct values into the relationship for resistors in parallel.
		(ii)	The ammeter reading decreases. The total resistance has increased.	(1) (1)	2	Increasing the resistance of any resistor connected in parallel will increase the combined parallel resistance of the circuit. Using $I = \dfrac{V}{R_T}$ means that the current will decrease.
					(9)	
8	a)		$E_h = cm\Delta T$ $= 4180 \times 250 \times 10 \cdot 3$ $= 1 \cdot 1 \times 10^7\,\text{J}$	(1) (1) (1)	3	Temperature change, $\Delta T = (25 \cdot 8 - 15 \cdot 5) = 10 \cdot 3\,°\text{C}$, must be calculated first. Then use $E_h = cm\Delta T$ to calculate the energy transferred into heat energy.
	b)		$P = \dfrac{E}{t}$ $350 = \dfrac{1 \cdot 1 \times 10^7}{t}$ $t = 3 \cdot 1 \times 10^4\,\text{s} = 8 \cdot 6$ hours	(1) (1) (1)	3	Use the value for heat energy calculated in part a). The question requires the answer, in seconds, to be converted into hours: (t in hours $= \dfrac{3 \cdot 1 \times 10^4}{60 \times 60} = 8.6$ hours).
					(6)	
9	a)		The water will start to boil. Mass of water in the container will reduce as the steam escapes.		1	This is a standard type of experiment to measure the specific latent heat of vaporisation of water. As the water boils, some of the water is vaporised and this steam escapes and the mass of water remaining reduces.

Question		Answer	Max. mark	Commentary with hints and tips
	b)	Mass of water vaporised = 1710 − 1460 = 250 g = 0·25 kg Energy used to vaporise water = $(7·52 − 1·62) \times 10^5 = 5·9 \times 10^5$ J (1) $E_h = ml$ (1) $5·9 \times 10^5 = 0·25 \times l$ (1) $l = 2·36 \times 10^6$ J kg^{-1} (1)	4	First, calculate the mass of water vaporised, then calculate the energy used during vaporisation. Then use the relationship for specific latent heat to calculate the value for the specific latent heat of vaporisation of water. Remember to include the correct units in the answer (J kg^{-1}). Hint: the units for the specific latent heat of vaporisation of water can be found in the Data sheet.
	c)	Insulating the metal container should improve the accuracy.	1	Insulating the metal container would mean that less heat energy would be lost to the surroundings. This means that less heat energy would be supplied to cause the same mass to be vaporised, which would lead to a smaller calculated value for the specific latent heat.
			(6)	
10	a)	$20\,°C = 293$ K $\dfrac{p_1 \times V_1}{T_1} = \dfrac{p_2 \times V_2}{T_2}$ (1) $= \dfrac{1·0 \times 10^5 \times 5·0 \times 10^{-4}}{293}$ (1) $= \dfrac{5·0 \times 10^6 \times 2·0 \times 10^{-5}}{T_2}$ $T_2 = 586$ K (1)	3	The temperature before compression must be converted into Kelvin before use in the relationship. It is not necessary to convert the final temperature back into °C unless the question specifically asks for this.
	b) (i)	Pressure difference = $7·0 \times 10^6 − 1·0 \times 10^5$ $= 6·9 \times 10^6$ Pa (1) $P = \dfrac{F}{A}$ (1) $7·0 \times 10^6 = \dfrac{F}{6·4 \times 10^{-3}}$ (1) $F = 4·48 \times 10^4$ N	3	First, calculate the pressure difference between inside and outside the cylinder. Then write $P = \dfrac{F}{A}$. Then substitute values into the relationship. Always write down the final answer in these 'show' type of questions.
	(ii)	The volume of the air increases. There are fewer collisions between the air particles and the piston walls because the air particles must travel further between collisions. Fewer collisions means the force on the piston walls is reduced, so pressure reduces.	1	It is important to mention: volume increasing, fewer collisions so reduced force on cylinder walls, resulting in reduced pressure.
			(7)	
11	a) (i)	$T = \dfrac{1}{f}$ (1) $= \dfrac{1}{0·25}$ (1) $= 4$ s (1)	3	Remember that when the frequency is in Hz, the answer for T is in seconds.

Question			Answer	Max. mark	Commentary with hints and tips
		(ii)	$v = f\lambda$ (1) $= 0{\cdot}25 \times 4$ (1) $= 1\,\mathrm{m\,s^{-1}}$ (1)	3	Remember to include the units for speed in the final answer.
		(iii)	 Wave direction → Wave machine	2	The diffracted waves after the barrier should: ▸ have the same wavelength as waves before the barrier (1) ▸ be circular and reach into the space beyond the barrier. (1)
	b)			3	Marks for the drawing are awarded as follows: ▸ 1 mark for showing the light ray changing direction at the air/water boundary. ▸ 1 mark for the angle of refraction in water being less than the angle of refraction in air. ▸ 1 mark for the normal being correctly drawn at the point where the light ray meets the air/water boundary. It is good practice to use a ruler when drawing light rays.
				(11)	
12	a)	(i)	infrared	1	The order of the bands of the electromagnetic spectrum should be memorised, including the direction of increasing wavelength and frequency of the bands.
		(ii)	photographic film or digital alternative (CCD)	1	Other detectors of X-rays include: Geiger–Müller tube, scintillation counter.
		(iii)	Gamma radiation has the highest frequency.	1	Gamma rays have the highest frequency and the shortest wavelength of the bands of the electromagnetic spectrum.
	b)		$d = vt$ (1) $2{\cdot}4 \times 10^{22} = 3 \times 10^8 \times t$ (1) $t = 8{\cdot}0 \times 10^{13}\,s$ $no.\,of\ light\ years = \dfrac{total\ time\ in\ seconds}{number\ of\ seconds\ in\ one\ light\ year}$ $\begin{array}{l}number\ of \\ light\ years\end{array} = \dfrac{8{\cdot}0 \times 10^{13}}{365{\cdot}25 \times 24 \times 60 \times 60}$ $= 2{\cdot}5 \times 10^6$ light years (1)	3	First, use $d = vt$ to calculate the time taken, in seconds, for light to travel $2{\cdot}4 \times 10^{22}$ m, using $v = 3 \times 10^8$ ms^{-1}. Then calculate the number of light years equivalent to this time: $no.\,of\ light\ years = \dfrac{total\ time\ in\ seconds}{number\ of\ seconds\ in\ one\ light\ year}$ (the number of seconds in one year = $365{\cdot}25 \times 24 \times 60 \times 60$)

Question			Answer	Max. mark	Commentary with hints and tips
	c)	(i)	Long wave signals diffract more around the Earth's curvature than short waves.	1	Wave diffraction is greater for longer wavelengths than short wavelengths, so longer wavelength radio waves from a land-based transmitter would diffract more around the curve of the Earth to reach ships that are far out on the sea.
		(ii)	$v = \dfrac{d}{t}$ (1) $3 \times 10^8 = \dfrac{d}{0 \cdot 0046}$ (1) $d = 1 \cdot 38 \times 10^6\,\text{m}$ (1)	3	Remember for this calculation that the speed of radio waves in air is $3 \times 10^8\,\text{m s}^{-1}$.
				(10)	
13			**Sample answer** ▶ Nuclear radiation is used in treatment of cancer patients. For example, gamma rays are directed into the tumour site from different angles to destroy cancer cells, but to minimise the damage to healthy cells. ▶ A radioactive 'tracer' is a liquid that contains a radioactive isotope which emits gamma radiation. The tracer is injected into the body and can be used to check the function of various organs of the body. For example, for a bone scan the tracer travels through the blood and collects in the bones. The radiation emitted by the tracer is detected by a gamma camera. The images obtained are analysed by computer, then displayed on screens to allow doctors to identify any problems. ▶ X-rays are used to get images of the body. X-rays travel through the body and photographic film or image intensifiers produce images of bones, tissue, etc. ▶ Infrared rays are used in the treatment of muscle injuries. ▶ Radio waves are used in the treatment of certain cancerous tumours. A probe which produces high-energy radio waves is directed into the tumour. The radio waves produce a current in the cells which heats them until they are destroyed.	3	This is an open-ended question: a variety of physics statements and descriptions can be used to answer this question. Marks are awarded on the basis of whether the answer overall demonstrates 'no' (0 marks), 'limited' (1 mark), 'reasonable' (2 marks) or 'good' (3 marks) understanding. 3 marks would be awarded to an answer which demonstrates a good understanding of the physics involved. The answer would show a good comprehension of the physics of the situation, provided in a logically correct sequence. (This type of answer might include a statement of the principles involved, a relationship or an equation, and the application of these to respond to the problem.) Also, note that the open-ended type of question is worth 3 marks. It is important not to spend too much time answering this question (3–4 minutes is the average time in the paper for 3 marks). A guide would be to take perhaps 3–5 minutes to complete this type of question. Only use more time than this if you have completed and checked all of the remaining questions in the paper.
				(3)	

Question			Answer	Max. mark	Commentary with hints and tips
14	a)		The lead box absorbs any background radiation.	1	Since background radiation is absorbed, the detected radiation is received only from the radioactive source.
	b)	(i)	Half-life is the time for the activity of a radioactive substance to reduce to half of its original value.	1	The definition of half-life is a standard question and should be memorised.
		(ii)	Activity is 90 kBq at time 0; activity is 45 kBq at time 9 hours. The half-life is 9 hours.	1	It is convenient to start with 90 kBq which is at time zero, then find the time on the graph for half (45 kBq); this allows the time to be read directly (9 hours).
	c)	(i)	The radioisotope emits radiation. (1) Large container prevents carriers from close contact with radioisotope. (1)	2	A safety precaution is to keep a distance between the radioisotope and the person carrying the box.
		(ii)	7 a.m. 5 July to 11 a.m. 6 July = 28 hours $800 \rightarrow 400 \rightarrow 200 \rightarrow 100 \rightarrow 3$ half-lives $\equiv 24$ hours so below 100 kBq in 28 hours (1) so radioisotope should not be used. (1)	2	First, calculate the time interval between delivery and the proposed use of the radioisotope. After three half-lives (24 hours) the activity is 100 kBq, so after 28 hours the activity would be *below* 100 kBq, so the radioisotope should not be used.
				(7)	
				[110]	

Practice Paper 2

Section 1

Question	Answer	Max. mark	Commentary with hints and tips
1	D	1	The net horizontal force is 10kN East and the net vertical force is 10kN South, so the resultant force must be in the direction between East and South.
2	E	1	Displacement = area under velocity–time graph $= \frac{1}{2} \times 2 \cdot 5 \times (-8) + \frac{1}{2} \times 2 \cdot 5 \times (+8)$ $= -10 + 10 = 0 \, \text{m}$ Acceleration $= \frac{v-u}{t} = \frac{8-(-8)}{5} = 3 \cdot 2 \, \text{m s}^{-2}$
3	B	1	First, calculate u and v: $u = \frac{d}{t} = \frac{0 \cdot 06}{0 \cdot 15} = 0 \cdot 4 \, \text{m s}^{-1}$, $v = \frac{d}{t} = \frac{0 \cdot 06}{0 \cdot 04} = 1 \cdot 5 \, \text{m s}^{-1}$ Then use $a = \frac{v-u}{t} = \frac{1 \cdot 5 - 0 \cdot 4}{0 \cdot 5} = 2 \cdot 2 \, \text{m s}^{-2}$.
4	D	1	Sloping part of a graph represents the deceleration, $t = 6 \, s$, $u = 55 \, \text{m s}^{-1}$, $v = 15$. Then use $a = \frac{v-u}{t} = \frac{15-55}{6} = -6 \cdot 7 \, \text{m s}^{-2}$. Take care to use u and v in the correct order the (negative sign represents deceleration).
5	B	1	$F_{un} = 4 - 2 = 2 \, \text{N}$ to the left, $a = \frac{F}{m} = \frac{-2}{4} = -0 \cdot 5 \, \text{m s}^{-2}$, i.e. a deceleration after 2 seconds, final velocity $v = u + at = 4 + (-0 \cdot 5 \times 2) = 3 \, \text{m s}^{-1}$.
6	C	1	At 350km, $g = 8 \cdot 5 \, \text{m s}^{-2}$, $W = mg = 90 \times 8 \cdot 5 = 765 \, \text{N}$.
7	A	1	The loss of potential energy is calculated using $E_p = mgh$ where $h = 0 \cdot 5 \, \text{m}$ and $g = 9 \cdot 8 \, \text{N kg}^{-1}$. Note that the length of the track and the frictional force are not required in this calculation.
8	D	1	Each ball takes the same time to reach the floor, because they are released from the same height. Different horizontal distances travelled mean that the horizontal velocities are different. Both balls fall for the same time from rest, so have the same final vertical velocities.
9	D	1	In a 'gravity assist' fly-by of a spacecraft with a planet which is in orbit around the Sun, some of the planet's kinetic energy is transferred to the spacecraft, increasing its speed in this case.
10	A	1	Weight of module on Moon $W = mg = 5000 \times 1 \cdot 6 = 8000 \, \text{N}$ $F_{un} = 15\,000 - \text{weight} = 15\,000 - 8000 = 7000 \, \text{N}$ $a = \frac{F}{m} = \frac{7000}{5000}$ $= 1 \cdot 4 \, \text{m s}^{-2}$
11	A	1	Rocket engines exert a downward force on exhaust gases so the reaction force is exhaust gases exerting an upward force on the rocket engines.
12	C	1	Use the Moon's gravitational field strength ($1 \cdot 6 \, \text{N kg}^{-1}$ from the Data sheet) to calculate the mass of the rover vehicle: $W = mg$, $m \frac{W}{g} = \frac{1360}{1.6} = 850 \, \text{kg}$ Then use Earth's gravitational field strength to calculate the Rover vehicle's weight on Earth: $W = mg = 850 \times 9 \cdot 8 = 8330 \, \text{N}$

Question	Answer	Max. mark	Commentary with hints and tips
13	E	1	The line spectra of hydrogen and nitrogen line up exactly with lines appearing in the distant star's spectrum. Hint: use a ruler to find which element has all of its lines exactly matching the position of the lines in the star spectrum.
14	D	1	First, calculate the current in the device with $P = IV$, using $V = 230\,V$ and $P = 9200\,W$ (from rating plate), then use $Q = It$, converting 5 minutes into seconds.
15	B	1	Use $V = IR$ where $V = 6\,V$ and $I = 0{\cdot}5\,A$ to calculate total resistance in the circuit, then subtract $5\,\Omega$ from the total resistance to obtain resistance of R.
16	E	1	To measure R, the current in the circuit, and the voltage directly across the resistor, must be measured.
17	C	1	Light-emitting diode: The symbols for cell, battery, lamp, switch, resistor, voltmeter, ammeter, LED, motor, microphone, loudspeaker, photovoltaic cell, fuse, diode, capacitor, thermistor, LDR, relay and transistor should be memorised.
18	C	1	Use $\dfrac{1}{R_T} = \dfrac{1}{R_1} + \dfrac{1}{R_2} + \dfrac{1}{R_3}$ to determine the resistance of resistors in parallel, and then add this answer to the $4\,\Omega$ resistor in series to get total resistance.
19	E	1	The resistors are connected in series. The power developed in each resistor is calculated using $P = I^2R$. So the greatest value of resistor develops the greatest power.
20	B	1	Temperature change, $\Delta T = 70 - (-15) = 85\,°C$, is the same in K as it is in °C.
21	C	1	The temperature of the air inside the syringe remains constant. This means that the average speed of the molecules does not change, and also that the air molecules do not strike the walls inside the syringe with greater force. However, because of the reduced volume, the air molecules collide with the walls inside the syringe more often.
22	D	1	Since $f = \dfrac{v}{\lambda}$, the smallest wavelength will produce the highest frequency. $f = \dfrac{v}{\lambda} = \dfrac{340}{0{\cdot}04} = 8500\,Hz = 8{\cdot}5\,kHz$
23	E	1	When moving from air into another medium, the speed and wavelength of light always change.
24	A	1	When light travels from glass into air or air to glass, the angle between the normal and the light ray is always greater in air than in glass.
25	B	1	Use $\dot{H} = \dfrac{H}{t}$; change $0{\cdot}08\,\mu Sv$ into $8{\cdot}0 \times 10^{-8}$; keep time in hours.

Section 2

Question		Answer		Max. mark	Commentary with hints and tips
1	a)	Negative		1	Particle must have negative charge because, due to the electric field, it experiences an upward force towards the positive metal plate.
	b)	$W = mg$ $\quad = 5\cdot4 \times 10^{-8} \times 9\cdot8$ $\quad = 5\cdot3 \times 10^{-7} \mathrm{N}$	(1) (1) (1)	3	Remember to state the correct units for weight, newtons, N.
	c)	Particle moves at constant velocity. Forces are now balanced.	(1) (1)	2	After 3 seconds, the horizontal line on the graph means constant velocity, so the forces on the particle are balanced.
				(6)	
2		**Sample answer** ▶ On the motorway, driving tends to be for long journeys at higher constant speeds than in town. This means that the engine uses energy only to overcome frictional forces. ▶ In town driving, there are many occasions where the car must slow down and then accelerate. In addition to the energy required to maintain constant speed, additional energy to produce the unbalanced force required for acceleration is needed. ▶ On the occasions when cars are in long queues or traffic jams, energy is used up while the car engine idles. This is possible on motorways and in towns. ▶ The car design also affects how much energy the car uses; if the car is streamlined, then less fuel will be required to overcome air friction.		3	This is an open-ended question: a variety of physics arguments can be used to answer this question. Marks are awarded on the basis of whether the answer overall demonstrates 'no' (0 marks), 'limited' (1 mark), 'reasonable' (2 marks) or 'good' (3 marks) understanding. 3 marks would be awarded to an answer which demonstrates a good understanding of the physics involved. The answer would show a good comprehension of the physics of the situation, provided in a logically correct sequence. (This type of answer might include a statement of the principles involved, a relationship or an equation, and the application of these to respond to the problem.) Also, note that the open-ended type of question is worth 3 marks. It is important not to spend too much time answering this question (3–4 minutes is the average time in the paper for 3 marks). A guide would be to take perhaps 3–5 minutes to complete this type of question. Only use more time than this if you have completed and checked all of the remaining questions in the paper.
				(3)	

Question			Answer	Max. mark	Commentary with hints and tips
3	a)	(i)	50 N	1	Care is required when identifying the correct point for the force on the graph corresponding to 0·35 m.
		(ii)	$E_w = Fd$ (1) $= 50 \times 0·35$ (1) $E_w = 17·5$ J (1)	3	Use the answer determined in part **a) (i)** as the distance, d to be used in the $E_w = Fd$ equation.
	b)	(i)	$s = vt$ (1) $30 = 40 \times t$ (1) $t = 0·75$ s (1)	3	The horizontal velocity, v, is constant because air resistance is ignored.
		(ii)	$E_k = \frac{1}{2} mv^2$ (1) $= \frac{1}{2} \times 0·06 \times 40^2$ (1) $E_k = 48$ J (1)	3	Take care when using $E_k = \frac{1}{2} mv^2$ to square the value of the speed when calculating the answer.
		(iii)	(A) $v^2 = 40^2 + 7·35^2$ (1) $v = 40·7$ m s^{-1} (1) resultant velocity magnitude is 40·7 m s^{-1} (B) $\tan x = \dfrac{7·35}{40}$ (1) $x = 10·4°$ Resultant velocity is 40·7 m s^{-1} at 10·4° below horizontal. (1)	4	The answer can be obtained by drawing a scale diagram; marks are awarded as follows: ▸ 1 mark for drawing a diagram to a reasonable scale, and for lines drawn with correct length and angle ▸ 1 mark for adding the horizontal velocity and final vertical velocity vectors and correctly showing resultant direction (arrow needed) ▸ 1 mark for resultant $v = 40·7$ ms^{-1} ▸ 1 mark for direction of 10·7° below horizontal.
				(14)	
4	a)		Sagittarius A	1	It is important to read the passage carefully, as the questions asked require information from the passage.
	b)		Approximately 14 billion years ago	1	Although this time is not mentioned, the passage states that 'scientists think the supermassive black holes formed when the universe began'.
	c)		$d = vt$ (1) $d = 3 \times 10^8 \times 27\,600 \times 365·25 \times 24 \times 60 \times 60$ (1) $d = 2·61 \times 10^{20}$ m (1)	3	Use $d = vt$ to calculate the distance in metres, from the centre of the Milky Way to the Earth. The speed of light $(3 \times 10^8$ m s$^{-1})$ is multiplied by the time for light to travel the 27 600 light years (i.e. 27 600 × 365·25 × 24 × 60 × 60 seconds). (the number of seconds in one year = 365·25 × 24 × 60 × 60) Note that 365·25 days is the average number of days in one year.
	d)		Geiger–Müller tube	1	Other gamma ray detectors include: scintillation counter.
				(6)	

Question			Answer		Max. mark	Commentary with hints and tips
5	a)	(i)	1. Chandra 2. IRIS 3. WISE (2 marks for all correctly identified; 1 mark for one correctly identified.)		2	The order of the bands of the electromagnetic spectrum should be memorised, and the order of increasing wavelength or frequency of these bands.
		(ii)	1. WISE 2. IRIS 3. Chandra (2 marks for all correctly identified; 1 mark for one correctly identified.)		2	Remember, the relationship between the altitude of a satellite and its period is: the greater the altitude, the greater the period of the satellite.
	b)		$v = f\lambda$ $3 \times 10^8 = f \times 2{\cdot}5 \times 10^{-7}$ $f = 1{\cdot}2 \times 10^{15}\,Hz$	(1) (1) (1)	3	Remember, all waves in the electromagnetic spectrum, including ultraviolet waves, travel at the speed of light, $3 \times 10^8\,m\,s^{-1}$ in a vacuum (or air). This value is found in the Data sheet.
	c)		$d = vt$ $105\,000 \times 10^3 = 3 \times 10^8 \times t$ $t = 0{\cdot}35\,s$	(1) (1) (1)	3	Use the altitude given in the table for the distance travelled by the radio signal. Remember, radio signals travel at the speed of light, $3 \times 10^8\,m\,s^{-1}$ in a vacuum (or air).
	d)		Radiation from some bands of the electromagnetic spectrum is absorbed by the Earth's atmosphere, so some telescopes need to be above the atmosphere to detect this radiation.		1	It is not necessary, in this answer, to name the bands of radiation which are absorbed by the Earth's atmosphere.
					(11)	
6	a)		Adjust variable resistor.		1	This alters total resistance in the circuit, altering the current in the circuit.
	b)	(i)	 Suitable scales, labels and units (1) All points plotted accurately to ± half a division (1) Best-fit curve (1)		3	Take care to choose suitable scales which will produce a graph of reasonable size on the grid provided (remember that a spare grid is always provided at the end of the question paper if you need to redraw your graph). The labels for each axis should be clearly identified with the correct unit supplied. The points must be plotted accurately. The final curve should be 'best fit', i.e. smoothly drawn through as many points as possible (but never simply join the plotted points to produce a 'zig-zag' effect).

Question			Answer		Max. mark	Commentary with hints and tips
		(ii)	$1{\cdot}2 \pm 0{\cdot}1$ V		1	Use a ruler along the graph to obtain an accurate reading. For the reading from the graph you would be allowed an answer of any value between 1·1 and 1·3 V.
		(iii)	Resistance increases as current increases (1) When $I = 0{\cdot}44$A, $V = 1{\cdot}6$V, so $R = \frac{V}{I} = 3{\cdot}6\,\Omega$ (1) When $I = 1{\cdot}0$A, $V = 9{\cdot}6$V, so $R = \frac{V}{I} = 9{\cdot}6\,\Omega$ (1)		3	An explanation is required, so choose values for voltage and current at different parts of the graph to calculate and show the lamp's resistance at these points.
					(8)	
7	a)	(i)	Lamps must operate at 230 V. In parallel, each lamp receives the supply voltage (230 V).		1	For each lamp to receive 230 V, they must be connected in parallel with the supply.
		(ii)	$P = \frac{V^2}{R}$ (1) $50 = \frac{230^2}{R}$ (1) $R = 1058\,\Omega = 1100\,\Omega$ (1)		3	Select the relationship $P = \frac{V^2}{R}$ to find resistance because the power and operating voltage of the lamp are given.
	b)		Total power $= 3 \times 50 = 150$ W (1) $P = IV$ $150 = I \times 230$ (1) $I = 0{\cdot}65$ A (1)		3	To calculate total current, use the total power of all three lamps.
					(7)	
8	a)		$E_h = cm\Delta T$ (1) $\quad = 4180 \times 0{\cdot}6 \times 75$ (1) $E_h = 188\,100$ J $= 190\,000$ J (1)		3	Temperature change, $\Delta T = 100 - 25 = 75\,^\circ$C, must be calculated first. Then use $E_h = cm\Delta T$ to calculate the energy required.
	b)	(i)	$P = \frac{E}{t}$ (1) $800 = \frac{E}{12 \times 60}$ (1) $E = 576\,000$ J (1)		3	Remember, time must be converted into seconds for this relationship.
		(ii)	$E_h = ml_v$ (1) $576\,000 = m \times 22{\cdot}6 \times 10^5$ (1) $m = 0{\cdot}25$ kg (1)		3	Use the energy calculated in b) (ii) in this relationship. The value for the specific latent heat of vaporisation, l_v, is found in the Data sheet.
		(iii)	Some of the heat energy supplied will be lost to the surroundings and so less will be available to convert water into steam.		1	There is always energy lost to the surroundings whenever a substance is heated by an appliance. This heat energy loss from the steamer to the surroundings means that there will be less energy available to convert water into steam.
					(10)	

Question		Answer		Max. mark	Commentary with hints and tips
9	a)	Pressure is the force per unit area exerted on a surface.		1	There are several definitions required in the National 5 Physics specification. Make sure that you know these definitions as you may be asked about them in the exam.
	b)	$p = \rho g h$ <div style="text-align:right">(1)</div>$= 1025 \times 9 \cdot 8 \times 24$ <div style="text-align:right">(1)</div>$= 2 \cdot 4 \times 10^5 \, \text{Pa}$ <div style="text-align:right">(1)</div>		3	Use the relationship given in the passage carefully and make sure that you use the correct numerical information from the passage. Remember to give the correct unit in the final answer.
				(4)	
10	a)	$T_1 = 10\,°\text{C} = 283\,\text{K}, \; T_2 = -40\,°\text{C} = 233\,\text{K}, \; p_1 = 1 \cdot 01 \times 10^5$ $\dfrac{p_1}{T_1} = \dfrac{p_2}{T_2}$ <div style="text-align:right">(1)</div> $\dfrac{1 \cdot 01 \times 10^5}{283} = \dfrac{p_2}{233}$ <div style="text-align:right">(1)</div> $p_2 = 8 \cdot 3 \times 10^4 \, \text{Pa}$ <div style="text-align:right">(1)</div>		3	Temperature must be converted from °C into Kelvin by adding 273. Use $\dfrac{p_1}{T_1} = \dfrac{p_2}{T_2}$ to calculate p_2.
	b)	(i)	$\Delta p = 1 \cdot 01 \times 10^5 - 8 \cdot 3 \times 10^4$ $= 1 \cdot 8 \times 10^4 \, \text{Pa}$	1	Use the answer to part **a)** in this calculation.
		(ii)	$P = \dfrac{F}{A}$ <div style="text-align:right">(1)</div> $1 \cdot 8 \times 10^4 = \dfrac{F}{0 \cdot 45}$ <div style="text-align:right">(1)</div> $F = 8100 \, \text{N}$ <div style="text-align:right">(1)</div>	3	The air pressure difference between inside and outside the container, calculated in part **b) (i)**, is used in this relationship.
		(iii)	Downwards on lid	1	The air pressure outside the container is greater than the pressure inside, so the resulting pressure is downwards on the lid.
				(8)	
11	a)	The direction of current from an alternating supply changes at the frequency of the supply. The current from a direct supply always moves in the same direction.		1	It is essential to clearly describe each type of current to identify the difference between them.
	b)	(i)	Amplitude $= 3 \times 2 \cdot 5 = 7 \cdot 5 \, \text{V}$	1	There are 3 centimetres from the centre to the top of the waveform, and the volts per centimetre setting is 2·5 volts.
		(ii)	Time T for 1 wave: $8 \times 0 \cdot 4 \times 10^{-3} = 3 \cdot 2 \times 10^{-3} \, \text{s}$	1	One complete wave occupies 8 boxes on the horizontal axis. The timescale setting is 0·4 milliseconds per centimetre, so total time for the wave (the period, T) is number of centimetres × timescale setting.
		(iii)	$T = \dfrac{1}{f}$ <div style="text-align:right">(1)</div> $3 \cdot 2 \times 10^{-3} = \dfrac{1}{f}$ <div style="text-align:right">(1)</div> $f = 312 \cdot 5 \, \text{Hz} = 310 \, \text{Hz}$ <div style="text-align:right">(1)</div>	3	Use the value for T calculated in part **b) (ii)**.

Question			Answer	Max. mark	Commentary with hints and tips
	c)	(i)	W only	1	**d.c. circuit:** In circuit A, the LED conducts current from the negative terminal to the positive terminal in W only. The orientation of LEDs Y and Z are the wrong way for conduction and so prevent current in the remaining branches.
		(ii)	W and Z	1	**a.c. circuit:** In an alternating current circuit, W and Z will conduct for each separate direction of the alternating current, but X and Y connected together in series will prevent any conduction of current in them.
				(8)	
12			**Sample answer** ▸ There is more diffraction of sound than of light; sound waves have longer wavelength than light waves, so the sounds of the dog are likely to diffract towards the student around fences or trees. ▸ Light waves are less likely to diffract enough to illuminate the side of the field, and the light from the torch is directed into a narrow beam. ▸ The sounds of the dog may reflect off fences or trees to reach the student. Sound waves are more likely to travel in all directions from the dog, allowing the student to hear the sounds directly without requiring diffraction.	3	This is an open-ended question: a variety of physics arguments can be used to answer this question. Marks are awarded on the basis of whether the answer overall demonstrates 'no' (0 marks), 'limited' (1 mark), 'reasonable' (2 marks) or 'good' (3 marks) understanding. 3 marks would be awarded to an answer which demonstrates a good understanding of the physics involved. The answer would show a good comprehension of the physics of the situation, provided in a logically correct sequence. (This type of answer might include a statement of the principles involved, a relationship or an equation, and the application of these to respond to the problem.) Also, note that the open-ended type of question is worth 3 marks. It is important not to spend too much time answering this question (3–4 minutes is the average time in the paper for 3 marks). A guide would be to take perhaps 3–5 minutes to complete this type of question. Only use more time than this if you have completed and checked all of the remaining questions in the paper.
				(3)	

Question			Answer	Max. mark	Commentary with hints and tips
13	a)		Alpha radiation consists of helium nuclei emitted from unstable atoms when they decay.	1	Alpha particles consist of two protons and two neutrons combined, which is a helium nucleus and is positively charged.
	b)		Ionisation is when an atom loses an electron to become a positive ion.	1	The process of ionisation is when neutral atoms lose (or gain) electrons to become ions. This can occur when the atom is exposed to the different types of radiation.
	c)		Alpha radiation causes greatest ionisation of air, which is required between the electrodes. **(1)** Alpha radiation would be absorbed by the plastic housing which makes the detector safer for use. Beta radiation and gamma radiation may penetrate the casing to reach the exterior of the casing. **(1)**	2	Knowledge of the relative ionising effect and penetration of alpha, beta and gamma radiation is required.
	d)		1 mSv	1	Knowledge of equivalent dose rate and exposure safety limits for the public and for workers in the radiation industries in terms of annual effective equivalent dose is required.
	e)		Americium-241 432 years **(1)** A short half-life source would cause the detected current to change over a short period of time, causing a false alarm. **(1)**	2	Since the detector relies on a change of detected current when smoke is present, the short half-life source would cause the current to change without smoke present, setting off a false alarm.
	f)		$A = \dfrac{N}{t}$ **(1)** $= \dfrac{1{\cdot}68 \times 10^7}{7 \times 60}$ **(1)** $= 4{\cdot}0 \times 10^4 \text{ Bq}$ **(1)**	3	Remember to convert minutes (or hours) into seconds when using time.
				(10)	
14	a)		Neutron	1	A qualitative description of nuclear fission (and fusion) is required in National 5.
	b)	(i)	1000 → 500 (From graph: 0 → 200 years) so half-life is 200 years	1	It is convenient with this graph to start with 1000kBq, which is at the start of the time axis, then find the time on the graph for half (500kBq), which is at 200 years. This allows the half-life to be easily assessed from the graph.
		(ii)	1000 → 500 → 250 → 125 → 62·5 is four half-lives **(1)** So total time = 4 × 200 = 800 years **(1)**	2	Use the half-life calculated in **b)**. From the starting activity (1000kBq) halve this value until 62·5kBq is reached. Always state the final answer (including the unit, years).

Question		Answer		Max. mark	Commentary with hints and tips
c)	(i)	$D = \dfrac{E}{m}$ (1) $= \dfrac{850 \times 10^{-6}}{70}$ (1) $= 1\cdot2 \times 10^{-5}\,\text{Gy}$ (1)		3	The relationship $D = \dfrac{E}{m}$ does not require the nature of the ionising radiation to be taken into account. Only the energy absorbed and the mass of the absorbing tissue are required.
	(ii)	$H = Dw_r$ (1) $= 1\cdot2 \times 10^{-5} \times 20$ (1) $= 2\cdot4 \times 10^{-4}\,\text{Sv}$ (1)		3	Use the value for the absorbed dose, D, from c) (i). The relationship $H = Dw_r$ to determine the equivalent dose requires the radiation weighting factor for the ionising radiation, w_r, to be obtained from the table of radiation weighting factors in the Data sheet (in this case, $w_r = 20$ for alpha radiation). If the type of ionising radiation is not in this table, then the value for w_r will appear in the question.
	(iii)	Worker should wear protective clothing		1	This precaution is one safety measure when dealing with radiation.
d)		Nuclear fusion		1	The question is mainly about nuclear fission. Nuclear fusion is the other nuclear process involved in the generation of electricity.
				(12)	
				[110]	